**MTT Student Pacing Guide – Hawkes – Spring 2017** **Website: Learn.HawkesLearning.com**

VCCS Email Address: drh2373@email.vccs.edu

Name: HELTON,DUSTIN RAY

My Program of Study is Curricular - AAS IT Game Design Spec which requires Units 1-5 to prepare for Math mth 151.

I have already completed Units 1-4 (either with the placement test or in an MTE or MTT class).

I am enrolled in MTT 1 and I will be completing Units 5 this semester.

## Weekly Assignment Goals:

| | Unit 1 | Unit 2 | Unit 3 | Unit 4 | Unit 5 |
|---|---|---|---|---|---|
| **Week 1** | 1.1, 1.2 | 2.1, 2.2, 2.3, 2.4 | 3.1, 3.2, 3.3, 3.4 | 4.1, 4.2 | 5.1, 5.2, 5.3 |
| **Week 2** | 1.3, 1.4 | U2 Proctored Q, 2.5, 2.6, 2.7, 2.8 | 3.5, 3.6, 3.7, 3.8 | 4.3, 4.4, 4.5 | 5.4, 5.5 |
| **Week 3** | 1.5, 1.6 | 2.9, 2.10, 2.11 | 3.9, 3.10, 3.11, 3.12 | 4.6, 4.7 | 5.6, 5.7, 5.8 |
| **Week 4** | 1.7 | 2.12, 2.13, 2.14 | 3.13, 3.14, 3.15, 3.16 | 4.8, 4.9 | 5.9, 5.10, 5.11 |
| **Week 5** | Review, Exam | Review, Exam | Review, Exam | Review, Exam | Review, Exam |

| | Unit 6 | Unit 7 | Unit 8 | Unit 9 |
|---|---|---|---|---|
| **Week 1** | 6.1, 6.2, 6.3, 6.4 | 7.1 | 8.1, 8.2 | 9.1, 9.2 |
| **Week 2** | 6.5, 6.6, 6.7 | 7.2, 7.3 | 8.3, 8.4 | 9.3 |
| **Week 3** | 6.8, 6.9, 6.10, 6.11 | 7.4, 7.5 | 8.5, 8.6 | 9.4, 9.5 |
| **Week 4** | 6.12, 6.13, 6.14 | 7.6 | 8.7, 8.8 | 9.6 |
| **Week 5** | Review, Exam | Review, Exam | Review, Exam | Review, Exam |

AF855670

## Weekly Checklist:

| Unit | Week | Dates | Sections | Goal met? | Notes |
|---|---|---|---|---|---|
| | 1 | Jan 23 – 27 | | | |
| | 2 | Jan 30 – Feb 3 | | | |
| | 3 | Feb 6 – 10 | | | |
| | 4 | Feb 13 – 17 | | | |
| | 5 | Feb 20 – 24 | Exam | | |
| | 6 | Feb 27 – Mar 3 | | | |
| | 7 | Mar 13 – 17 | | | |
| | 8 | Mar 20 – 24 | | | |
| | 9 | Mar 27 – 31 | | | |
| | 10 | Apr 3 – 7 | Exam | | |
| | 11 | Apr 10 – 14 | | | |
| | 12 | Apr 17 – 21 | | | |
| | 13 | Apr 24 – 28 | | | |
| | 14 | May 1 – 5 | | | |
| | 15 | May 8 | Exam | | |

**Restart Dates**: February 24 and April 7.

Spring 2017

## MTT – Developmental Math – Syllabus – Hawkes

**Instructor Contact Info & Office Hours**: Listed in Blackboard and here.

This is a developmental math class. It is pass/fail (S/U) and will not affect your GPA and the credits will not count towards graduation. The number after the "MTT" in your course title refers to how many Units you are completing this semester, either 1, 2, or 3. In order for you to be successful in this course, you will need to plan on spending *at least 12 hours per week (including in-class time)* working on your math.

If you need help, **free math tutoring** is available to all NRCC students at Academic Assistance, located both in Dublin (Godbey 131) and at the Mall site (room 202). Online tutoring sessions are also available. They also offer several online tutorials in Blackboard under the NRCC Tutoring Services tab. Contact them at http://www.nr.edu/aa/ or 540.674.3664.

Make sure you know which Units you are completing this semester, so that you are sure to start in the correct one. See your individual Pacing Guide for this information, and make sure you agree with the information you are given (i.e. that you're in the correct program and starting in the correct Unit). If changes need to be made, it is your responsibility to notify your instructor right away.

**Required Materials**: *Foundations of Mathematics for Virginia*, published by Hawkes. (Guided Notebook and software access code bundle.) ISBN 978-1-941552-88-9. Materials must be purchased from NRCC bookstore as they are custom. (This bundle is good for all 9 Units, so if you have used it in a previous MTT class, you will use it again for this MTT class.) For this course, all your work will be done using Hawkes Learning Systems by logging in to Learn.HawkesLearning.com. You will also need to purchase a binder for your Guided Notebook, and have lined notebook paper. No calculator is allowed for Unit 1 and the first 4 sections of Unit 2, but a scientific calculator (TI-30XIIS) is allowed for the rest of Unit 2 and beyond. Graphing calculators are discouraged for this class.

### How to Proceed Through Your Course

Each Unit must be completed in 5 weeks. Time management will be your biggest ally during this course! Do not allow yourself to get more than a few days behind your Pacing Guide! If you fall behind the Pacing Guide, you will be in danger of failing the course since you won't be able to finish all your Units before the end of the semester. In addition, **IF YOU DO NOT FINISH EACH UNIT IN THE 5-WEEK BLOCK, YOU WILL BE REQUIRED TO RESTART THAT UNIT FROM THE BEGINNING**. *The restart dates for this semester are noted at the bottom of your Pacing Guide.* Of course, you are welcome to work ahead of the Pacing Guide; that provides a cushion for when life gets in the way. As soon as you complete your last Unit, you're done with the class and don't need to attend any more this semester!

You will start each Unit with a Unit Pre-Test. The Pre-Test allows you to demonstrate what you already know about the material in your Unit. Any section you demonstrate mastery in will not have to be completed. However, do not be discouraged if you do not know ANY, or very little, of the material in your Unit; that only means you are starting in the right place! You are not expected to know any of the material yet – that is why you are taking this class! **The password for the Pre-Test is *password*.**

After you finish the Unit Pre-Test, you are ready to get to work in your Unit. Each Unit has several sections, and each section involves the following steps:

1. "Learn" – this is where you learn the material covered in this section. Read through the pages, filling in your Guided Notebook as you go along. Work the examples on notebook paper, and store them in your Guided Notebook. *Filling in your GNB is a mandatory part of this class, and you must have a completed GNB before you will have your questions answered in class, or before the tutors at Academic Assistance will work with you.* You

are required to show your completed GNB to your instructor before taking your Unit Exam. There is also a video for each section, and a link to additional videos, under the "Watch" link in the top left corner of the page.

2. "Practice" – this is a set of problems for you to practice what you have learned. There are Learning Aids built in that can help you if you get stuck while you're working on a problem. You should work through every problem.
3. "Certify" – this is a set of problems just like those in Practice, but without the Learning Aids. You must score an 80% on this assignment to demonstrate mastery and move on to the next section. If you are working in Certify and miss too many questions, it will close and force you to go back into Practice. Once you've worked 90% of the Practice problems, you will be allowed to access Certify again. You may re-try Certify as many times as you need to until you earn an 80%, and then you will get full credit for that section and be allowed to move into your next section.

When you have completed all the sections in your Unit, you will be ready to take the Practice Exam for your Unit. After you submit this test, you can access a report that will tell you exactly what sections you need to review before taking your Unit Exam. Click on the sections (in pink) you need to work on in the bar graph, and you will have a customized set of practice problems based on what you missed on the Practice Exam. Do this for each section where you missed any problems before trying the Practice Exam again. You will not be allowed to take your Unit Exam until you've made at least an 80% on the Practice Exam. You may take the Practice Exam as many times as you want or need.

You will take your Unit Exam in class with your instructor (or at the Testing Center if you are a DE student). You will have 3 attempts to pass the Unit Exam; you must make a 75% or better on the Unit Exam to pass that Unit and move on to the next one. If you do not pass it on the first attempt, you will need to repeat the Practice Exam before you will be allowed to try the Unit Exam for a second time. You will also need to meet with your instructor to go over your failed Unit Exam and show her/him your completed Guided Notebook. If you do not pass the Unit Exam in three attempts, you will need to re-start the Unit from the beginning. Your instructor will give you the details if this happens to you. ***NEW**: If you are on track with your Pacing Guide each week, you can earn a 4th attempt at the Unit Exam if you need it.

There is no final exam for this course. When you complete your last Unit of the semester, you are finished with the course.

**Course Grade**

This is a pass/fail (S/U) course. To pass the course, you must complete **ALL** the Units you signed up for. If you don't finish them all, then you will fail the course. You will not have to repeat any Units you did complete, but you will need to enroll in another MTT course to finish up the rest of your Units. As soon as you have completed all the Units you have signed up for, you are finished with the course, and you will earn a grade of S.

The **Accountability in Student Learning Program (ASLP) Team** connects students to the resources they need in order to meet their educational goals. Connection Specialists and Career Coaches intentionally engage students who may be in need of academic advising, tutoring, or help in overcoming obstacles to academic success such as lack of books, computers or Internet access at home. NRCC and the ASLP program cannot solve every problem; however, they CAN connect you to various community resources that are available. The ASLP Team also creates opportunities for students with similar backgrounds, goals and/or challenges to come together to support each other. If you are in need of academic assistance or other types of support – or if you have an idea for supporting student success and you don't know where to go – let us know. To contact the ASLP office for any reason, including to find out who your connection specialist is, email aslp@nr.edu or call 540.674.3677 or drop by Godbey 53 (in Dublin) or room 103 at the Mall Site.

**MTT Q & A: Pretests and Unit Exams**

Place this document in your Guided Notebook and bring it to class with you every day. You will be asked periodically to refer to it for the information it contains.

1. What score do I need to pass the Unit Exam?

A score of 75% or better.

2. How many times may I take a Unit Exam?

You have 3 attempts to pass a Unit Exam. If you fail your first attempt, you must show your instructor your completed Guided Notebook and go over your failed Unit Exam with her/him.

3. How many times can I take the Pretest?

You may only access a Pretest one time, but you can pause it once.

4. Can I test out of a Unit with the Pretest?

Yes, if you take it in class (in one sitting) with no notes or help, and you make a 75% or better.

5. What happens on testing days in class?

The last day of class each 5-week block is Unit Exam testing day. You must have either already tested your Unit by then, or you must show up for class on time and **ready to take the Unit Exam**.

6. What does "**ready to take the Unit Exam**" mean?

This means you have completed all the sections (Certifications) for your Unit (and passed the Unit 2 Proctored Quiz if you are doing Unit 2), and made an 80% or better on the Practice Exam for your Unit BEFORE THE START OF THE CLASS PERIOD.

7. What happens if I am not ready to test my Unit on testing day?

Your Unit will restart and you will have to start it over from the Pretest. You will have the chance to test out of the Unit by taking the Pretest in class that day with no notes or help and making a 75% or better. If you pass the Pretest this way, you will be able to start your next Unit. If you do not pass the Pretest, then you will have to redo the Certifications for that Unit (and any proctored quiz) and the Practice Exam and be ready to test it on time on the next Unit Exam testing day. This may result in your not having enough time to complete all your Units and failing the course.

8. Can I take a Unit Exam early?

Absolutely! The sooner you test and pass a Unit, the more time you will have for any remaining Units, and/or the sooner you will be finished with this class.

9. What is the password for the Pretest?

password

10. What is the password for the Unit Exam?

Your instructor will type that in for you.

11. May I get help with the Certifications?

No. You must do them on your own. If you need help, you must go back to Practice, and then your instructor or a tutor can help you. You must also have completed the Learn section, and filled in your Guided Notebook, in order to receive help with Practice.

12. When are the Unit Exam testing days for this course?

For MWF classes, the testing days this semester are: **February 24 and April 7**.
For TR classes, the testing days this semester are: **February 23 and April 6**.

**Reminders**:

- All Unit Exams must be taken in class with no notes or help from other sources.
- Pretests must be taken in class, with no notes or help from other sources, if you are wanting to try to test out of a Unit.
- On the final day of class for the semester, all the Certifications for the Unit **AND** an 80% or better on the Practice Exam must be completed **BEFORE CLASS STARTS**.
- All cell phones must be put away while testing!

# Getting Started Checklist – Hawkes Learning Systems (HLS)

NRCC Developmental Mathematics
Spring 2017

1. Check your Pacing Guide to be sure you are enrolled in the correct MTT course (1, 2, or 3, depending on how many Units you need to complete) and to see which Units you are completing this semester. If you see any problems with this information, contact your instructor right away.

2. Read the Syllabus for your course (given to you by your instructor). Make sure you understand how the course works. Your instructor will have you sign and detach the last page and turn it in to him/her.

3. **NEW MTT students**, follow these steps: (Returning MTT students, skip to step 4.)

    a. Open your web browser and go to learn.hawkeslearning.com.
    b. Click on "Create an Account".
        i. If you have already purchased your materials for this course, enter your license number.
        ii. If you have not yet purchased your book for this class, click on "I want to request Temporary Access", and then click "Continue". The temporary access code is good for 20 days from the start of the semester. When that time is up, you will be asked to enter a valid access code, so please purchase your code and Guided Notebook from the bookstore as soon as possible.
    c. You will then be led through the account creation process. You must use your VCCS email address to create your account.
    My email address: ________________@email.vccs.edu Password: __________________
    d. In the drop-down menus, choose New River Community College, choose *Foundations of Mathematics for Virginia* as your product, select your instructor, and then select the section for the first Unit you will be completing this semester. Now click "Enroll". Follow the steps to verify your email address. Log in with your username and password, and you will see the Dashboard for your course.
    e. Click on your name in the top right corner to **watch the Training Video** found in the drop down menu under your name. This will show you how to navigate and understand the courseware.

4. **RETURNING MTT students**, follow these steps:

    a. Open your web browser and go to learn.hawkeslearning.com.
    b. Log in with your VCCS email address and the password you used when you created your account. (If you have forgotten your password, click on "Forgot password?" and follow the instructions.)
    c. Enroll yourself in the correct section: Click on Add Course and give it a second to load. Then click on the "Edit" button next to the Foundations of Mathematics for Virginia section. In the drop-down menus, select your instructor, and then the section for your first Unit (found on your Pacing Guide). Last, click on Save Changes. To return to your course, click on Dashboard.
    d. Click on your name in the top right corner to **watch the Training Video** found in the drop down menu under your name. This will show you how to navigate and understand the courseware.

5. Click on the "View Course" button on the right to bring up the full To-Do List. Click on your Unit Pretest to start it first. In the password box type *password* and click "OK". Do not worry about your score, just do your best! If you demonstrate mastery in any of the sections in your Unit, you will not have to complete that section(s). Click on "Back to Course".

6. Click on "To Do List", and click on your first section. Click "Learn" to go through the material covered in the lesson. As you work your way through Learn, fill in your Guided Notebook. You should also work out the examples that are presented in Learn; you will want to do this on blank notebook paper and then file it in your Guided Notebook for that section. There is also a video to watch that will help explain the concepts in the section; to access it, click on "Watch" on the left side of the screen. When you are finished, click on "End Learn".

7. Click on "Practice" to complete the Practice problems in the lesson, taking advantage of the "Tutor", "Explain Error", and "Try Similar" options to further your understanding of the material and any mistakes you have made. Complete all the Practice problems.

8. Click on "Certify". These are the same kinds of problems you worked in Practice, but you don't have access to the learning aids here. These are mastery-based assignments, so you must get around 80% correct to get credit for the assignment. There is a mastery meter at the top which will let you know how you are doing. You can also see how many lives you have in the upper right hand corner. If you answer too many questions incorrectly, and run out of lives, you will be taken out of the Certify session and placed back into Practice, and given an opportunity to work on some of the problems you struggled with. When you are ready, you can try Certify again. As long as you eventually reach the ~80% mastery level, you will receive full credit for the homework. Once you have passed your Certify, you are ready to start the next section of your Unit.

9. Keep on working! Now that you have finished your first section, continue through steps 6 – 8 for each of the other sections in your Unit. You should expect to spend a total of about 12 hours per week working on course materials in order to be successful (that's about 2 hours per day, 6 days per week!).

10. Are you motivated? By working ahead of schedule and finishing your Unit early, you can start the next Unit right away and potentially finish the course early! But do not fall behind your Pacing Guide: <u>if you do not finish your Unit before the restart date, you will have to restart that Unit from the beginning</u>. The restart dates for this semester are listed on the bottom of your Pacing Guide. So be sure to meet your weekly deadlines stated on your Pacing Guide to avoid having to redo anything!

**Technical Troubleshooting:**

- Having trouble inside the program? **Contact Hawkes Tech Support at 1-800-426-9538 (M – F, 8:30 AM – 10 PM), or through 24/7 live chat at <u>www.HawkesLearning.com/chat</u>. You can also email them at <u>support@hawkeslearning.com</u>.**

# Foundations of Mathematics for Virginia

## GUIDED NOTEBOOK

Jennifer Comer

HAWKES LEARNING

**A division of Quant Systems, Inc.**
546 Long Point Road, Mount Pleasant, SC 29464



Printed in the United States of America

# Table of Contents

## 3a. Operations with Real Numbers



## 3b. Introduction to Algebra and Solving Equations



## 3c. Geometry



## 4. First Degree Equations and Inequalities in One Variable

## 6b. Factoring Polynomials and Solving Quadratic Equations



## 7. Rational Expressions and Equations



## 8. Roots, Radicals, and Complex Numbers

# Exponents, Prime Numbers, and LCM

- The result of multiplication is called the ____________________.
- The numbers being multiplied are called the ____________________ of the product.
- The type of operation used when working with exponents is ____________________ multiplication.
- In the expression, $7^2$, the number that will be multiplied repeatedly is __________.
- In the equation, $5^4 = 625$ __________ is the base, __________ is the exponent, and __________ is the exponential expression.
- Using the exponential expression, $11^3$, circle the equivalent expression below:

$11 \cdot 11 \cdot 11$ or $3 \cdot 3 \cdot 3 \cdot 3 \cdot 3 \cdot 3 \cdot 3 \cdot 3 \cdot 3 \cdot 3 \cdot 3$

- The exponent of 2 is represented by the word ____________________.
- The exponent of 3 is represented by the word ____________________.
- Exponents other than 2 and 3 can be read using the expression, "____________________ ____________________ (fill in the number) ____________________."
- A prime number is a counting number greater that _____ that has exactly _____ different __________.
- A prime number has the factors of __________ and __________.
- A composite number has __________ than __________ different factors.
- Circle all examples of prime numbers below:

5 13 21 19 100 102

- Circle all examples of composite numbers below:

2 7 18 25 49 103

- Steps to find the prime factorization of a composite number:

1. Factor the composite number into any two __________.
2. Factor each factor that is not __________ into __________ more factors.
3. Continue this process until all factors are __________.

- The prime factorization is the ____________________ of all the ___________ factors.

- A number is divisible:

By 2 if the units digit is _____, _____, _____, _____, or _____

By 3 if the sum of the digits is divisible by _____

By 5 if the units digit is _____ or _____

By 6 if the number is divisible by _____ and _____

By 9 if the sum of the digits is divisible by _____

By 10 if the units digit is _____

## Introduction to Fractions and Mixed Numbers

- The parts of a fraction are (shown below):

_____________ ————————→ $\frac{5}{8}$ ←———— _____________

- Fractions are used to indicate ____________________ of a whole.

- Use the picture below to write a fraction representing the shaded portion of the shape:

_________________

- The fraction $\frac{2}{7}$ represents _____ of _____ equal parts.

- Whole numbers can be thought of as fractions with a denominator of

____________________.

- Fraction notation represents the operation of ____________________.

- There are two rules to keep in mind when working with the value of zero in a fraction:

- For any nonzero value, b, $\frac{0}{b} = 0$. An example of this would be $\frac{0}{4} = 0$.

- For any value of a, $\frac{a}{0} =$ undefined. An example of this would be $\frac{6}{0} =$ undefined.

- Steps to multiply fractions:

  1. Multiply the ____________________.
  2. Multiply the ____________________.

- Multiply the following fraction: $\frac{2}{3}\cdot\frac{1}{5}=$ ________________ = ________

- To find an equivalent fraction, you need to ____________________ the numerator and the denominator by the ____________________ nonzero whole number.

- Find an equivalent fraction for $\frac{2}{3}=$ ________ by multiplying both the numerator and the denominator by 5.

- There are two steps to reducing fractions to lowest terms:

  1. Factor the ____________________ and ____________________ into prime factors.

  2. Use the fact that $\frac{k}{k}=1$ and divide out all of the ____________________ factors.

- What is the common factor that can be divided out of the fraction, $\frac{4}{12}$?

- What is $\frac{4}{12}$ in lowest terms?

- A mixed number is the sum of a ____________________ and a ______________________ fraction.

- To change a mixed number to an improper fraction, you need to:

  1. Multiply the whole number by the ____________________ of the proper fraction.

  2. Add the ____________________ of the proper fraction to this product.

  3. Write this ____________________ over the denominator of the fraction.

- Change $3\frac{1}{2}$ to an improper fraction:

  Multiply the whole number by the denominator: _____·_____ = _____

  - Add the numerator to the product from above: _____+_____= _____

  - Write this sum over the denominator: _____

## Ratios and Rates

- Two meanings for fractions are:
  - To indicate ____________________ of a whole.
  - To indicate ____________________.
- A ratio is a ____________________ of two quantities by

  ____________________.
- A ratio, 3 days to 4 nights, can be written in three ways:

  1.

  2.

  3.
- Ratios have the following characteristics:
  - Ratios can be ____________________.
  - Whenever units of the numbers in a ratio are the same, the ratio has

    ____________________ units, otherwise known as an ____________________ number.
  - When the numbers in a ratio have different units, then the numbers must be

    ____________________ to clarify what is being compared.
- A ____________________ is a statement that two ratios are equal.
- A proportion is ____________________ if the cross products are equal.

## Multiplication and Division with Fractions and Mixed Numbers

- To multiply with mixed numbers, you will need to change the mixed numbers to ____________________

  ____________________ and then multiply the fractions.

- You can multiply fractions and mixed numbers while reducing at the same time by using

  ____________________ factors.

- Multiply and reduce to lowest terms:

$$1\frac{2}{5}\cdot\frac{5}{8} = \_\_\_\_\_\_\_\_\_\_\_\_\_\_\_\_\_\_\_\_ = \_\_\_\_$$

- Remember that if all of the factors in the ____________________ or the ____________________ divide

  out, then the value of _____ must be used as a factor.

- The reciprocal of $\frac{a}{b}$ is $\frac{b}{a}$ (when $a$ and b do not equal 0).
  The product of a nonzero number and its reciprocal is always _____.

- Match the following values and its reciprocal:

| | |
|---|---|
| 15 | 2 |
| $\frac{7}{8}$ | $\frac{1}{15}$ |
| $\frac{1}{2}$ | $\frac{8}{7}$ |

- To divide fractions by any nonzero number, you will ____________________ by its reciprocal.

  For example, $\frac{1}{2}\div\frac{2}{5}=\frac{1}{2}\cdot\frac{\ }{\ }$ .

## Addition and Subtraction with Fractions

- To **add** fractions with the **same denominator**, you will need to:
  - ____________________ the numerators.
  - Keep the ____________________ denominator.
  - ____________________, if possible.
- To **add** fractions with **different denominators**, you will need to:
  - Find the least common ____________________ (LCD). Remember that the LCD is the least common multiple for the denominators.
  - Change each fraction into an ____________________ fraction with that denominator.
  - ____________________ the new fractions.
  - Reduce, if possible.
- A common error that occurs when adding fractions that needs to be avoided is ____________________ out across the addition sign.
- To **subtract** fractions with the **same denominator**, you will need to:
  - ____________________ the numerators.
  - Keep the ____________________ denominator.
  - Reduce, if possible.
- To **subtract** fractions with **different denominators**, you will need to:
  - Find the least common ____________________ (LCD).
  - Change each fraction into an ____________________ fraction with that denominator.
  - ____________________ the new fractions.
  - Reduce, if possible.

## Addition and Subtraction with Mixed Numbers

- Steps to adding or subtracting mixed numbers;

    1. Add or subtract the ____________________ parts, based on the operation in the problem.
        - Remember to get common ____________________ before you add the fractions.
        - You also need to ____________________ the fraction.
    2. Add or subtract the ____________________ numbers.
    3. Write the answer as a ____________________ number with a fraction part ____________________ than one.

Add the two mixed numbers in the example:

Add the mixed numbers: $3\frac{5}{7}+4\frac{2}{3}$

Get a common denominator for the fractions: $\frac{15}{21}+\frac{14}{21}$

Add the fractions: $\frac{29}{21}$

Change the fraction into a mixed number: $1\frac{8}{21}$

Add the whole numbers: $3+4+1$

Combine for a mixed number: $8\frac{8}{21}$

## Order of Operations with Fractions and Mixed Numbers

- To compare two fractions (to find which one is larger or smaller):
    - Find the least common ____________________.
    - Change each fraction to an ____________________ fraction with that ____________________.
    - Compare the ____________________.
- Rules for order of operations:
    - Simplify within ____________________ symbols, like parentheses, brackets, and braces, starting with the ____________________ most group.
    - Evaluate any numbers or expressions indicated by ________________.
    - Moving from ____________________ to ____________________, perform any multiplication or division in the order in which they appear.
    - Moving from ____________________ to ____________________, perform any addition or subtraction in the order in which they appear.

## Introduction to Decimal Numbers

- Decimal notation uses a ____________________ ____________________ system and a ____________________ point, with whole numbers written to the ____________________ and fractions written to the ____________________ of the decimal point.
- To read or write a decimal number:
  - ____________________ (or write) the whole number.
  - Read (or write) the word "____________________" in place of the decimal point.
  - Read (or write) the ____________________ part as a whole number. Then name the fraction with the name of the last ____________________ on the right. Add "th" to the end of the fraction place value.
  - Remember that if there is not a whole number, you can put a ____________________ to the left of the decimal point.
- Write the following mixed number as a decimal number:

  $3\frac{4}{100} =$ _____._____ _____

  and it would be read as _____ AND _____ ____________________
- When writing seventeen and 5 thousandths in decimal notation you would have ____________________ holders in the tenths and hundredths place values.

It would look like 17.005.

- When comparing decimals:
  - Moving ____________________ to ____________________, compare digits with the same ____________________ value.
  - When one compared digit is larger, the ________________________________ is larger.

- Compare the following values:

  5.789 Notice that the numbers are lined up, for easier comparison.
  5.754

  When moving from left to right, the digits are the same until the ____________________ place values. Those are going to be used to compare. Since the 8 is a larger value than 5, ____________________ is a larger value.

- Rules for rounding decimals:

  - Look at the digit to the ____________________ of the place of desired accuracy.
  - If this digit is 5 or ____________________, make the digit in the desired place of accuracy one larger and replace all digits to the right with zeros.

  All digits to the left remain unchanged unless a 9 is made one larger. This effectively changes the 9 to 10 which means the next digit to the left must be increased by 1.

  - If this digit is ____________________ than 5, leave the digit in the desired place of accuracy as it is and replace all digits to the right with zeros.

  All digits to the left remain unchanged.

  - Zeros **to the right of the place of accuracy** and to the right of the decimal point must be ____________________. In this way the place of accuracy is clearly understood. If a rounded number has a 0 in the desired place of accuracy, then that 0 remains.

Round 13.2687 to the nearest hundredth.

The digit in the hundredths is ____________________ .

The digit in the place value to the right of the hundredths is ____________________ .

Since that digit is greater than 5, the 6 in the hundredths place value changes to a 7.

The rounded value is _____._____ _____

## Addition and Subtraction with Decimal Numbers

- To add decimal numbers
    - Write the addends ______________________.
    - Keep the decimal points aligned ______________________.
    - Keep digits with the ______________________ position value aligned. (Zeros may be written in to help keep the digits aligned properly.)
    - Add, just as with whole numbers, keeping the decimal point in the sum ______________________ with the other decimal points.
- Find the sum: 2.1 + 7.83 + 8

  Be sure to align the place values before you begin to add. You can also write in zeros in missing decimal place values.

$$\begin{array}{r} 2.10 \\ 7.83 \\ \underline{+8.00} \end{array}$$

- To subtract decimal numbers:
    - Write the numbers ______________________.
    - Keep the decimal points aligned ______________________.
    - Keep digits with the ______________________ position value aligned. (Zeros may be written in to help keep the digits aligned properly.)
    - Subtract, just as with whole numbers, keeping the decimal point in the difference ______________________ with the other decimal points.
- Find the difference: 18.7 – 12.541

  Be sure to align the place values before you begin to subtract. You can also write in zeros in missing decimal place values.

$$\begin{array}{r} 18.700 \\ \underline{-12.541} \end{array}$$

- To estimate the sum or difference, you will:
    - Round each number to the place of the leftmost nonzero digit
    - Add or subtract the rounded values.

Estimate the difference: 65 – 12.43 – 24.762

65 rounds to 70

12.43 rounds to 10

24.762 rounds to 20

Now you can subtract: 70 – 10 – 20 = _____

## Multiplication with Decimal Numbers

- Multiplication of ____________________ is similar to multiplication of ____________________ numbers.
- There is no need to keep the decimal points lined up for ____________________.
- To multiply decimal numbers:
  - Multiply the two numbers as if they were ____________________ numbers.
  - Count the ____________________ number of places to the right of the decimal points in both numbers being multiplied.
  - Place the decimal point in the product so that the number of decimal places is the ____________________ as that found in Step 2.
- Multiply $17.451 \times 3.2$.

$$\begin{array}{r} 17.451 \\ \times \quad 3.2 \\ \hline \end{array}$$

Remember to count the decimal places in both values.

The decimal point will be _____ places into the product

(move from right to left)

- If the decimal point needs to be inserted further left in a value than there are digits, _____'s will need to be inserted.
- To multiply by a power of 10:
  - Count the number of _____'s in the power of 10.
  - Move the decimal point to the ____________________ the number of places equal to the number of 0s.
- Mental Math Quick Tips:
  - Multiplication by 10 moves the decimal point ____________________ place to the right.

- Multiplication by 100 moves the decimal point ____________________ places to the right.
- Multiplication by 1000 moves the decimal point ____________________ places to the right.
- The pattern keeps going….

## Division with Decimal Numbers

- The process of division with decimal numbers is similar to division with whole numbers. With whole numbers, we are concerned with the ____________________ but with decimal numbers, we are concerned with the ____________________ in the quotient.
- To divide decimal numbers:
    - Move the decimal point in the divisor to the ____________________ so that the divisor is a whole number.
    - Move the decimal point in the dividend the ____________________ number of places to the right.
    - Place the decimal point in the quotient directly ____________________ the new decimal point in the dividend.
    - Divide just as with ____________________ numbers. (0's may be added as needed to the dividend.)
- What to do if the remainder is not 0:
    - Decide first how many ____________________ ____________________ are to be in the quotient.
    - Divide until the ____________________ is one digit past the place of desired accuracy.
    - Using this last digit, round the ____________________ to the desired place of accuracy.
- Types of decimals:
    - Is eventually 0, the decimal number is said to be ____________________.

      Example: 3.5
    - If the remainder is not eventually 0, the decimal number is said to be ____________________.

      Example: 9.888888888…….
- To divide a decimal number by a power of 10:

  - Count the number of _____'s in the power of 10.
  - Move the decimal point to the ____________________ the number of places equal to the number of 0s.
- Mental Math Quick Tips:
  - Division by 10 moves the decimal point ____________________ place to the left.
  - Division by 100 moves the decimal point ____________________ places to the left.
  - Division by 1000 moves the decimal point ____________________ places to the left.
  - And the pattern keeps going….
- In order to estimate with division, ____________________ both the divisor and the dividend to the place of the last nonzero digit on the left and then ____________________ with these rounded values.

## Decimal Numbers and Fractions

- To change from decimal numbers to fractions:
  - A terminating decimal number can be written in fraction form by writing a fraction with the following:
    - a ____________________ that consists of the whole number formed by all the digits of the decimal number
    - a ____________________ that is the power of ten that names the position of the last digit on the right
- To change from a fraction to a decimal number:
  - To change a fraction to decimal form, we divide the numerator by the denominator.
    - If the remainder is eventually 0, the decimal number is said to be ____________________.
    - If the remainder is never 0, the decimal number is said to be ____________________.
- Nonterminating decimal numbers can be ____________________ or nonrepeating.
- A ____________________ repeating decimal (also called an infinite repeating decimal number) has a repeating pattern to its digits.
- Every fraction with a whole number numerator and nonzero denominator is either terminating or repeating. Such numbers are called ____________________ numbers.
- Sometimes, changing fractions to decimal form may involve ______________ the decimal form of a number and settling for an approximate answer. To have a more accurate answer, we may need to change the number from decimal form to ________________ form and then perform the operations.

## Decimal Numbers and Percents

- The word percent comes from the Latin *per centum*, meaning per

  ____________________.

- Percent means ____________________, or the ratio of a number to 100.
- The symbol ____________________ is called the percent sign. This sign has the same

  meaning as the ____________________ $\frac{1}{100}$.

- Changing fractions with denominators of 100 to percents:
- Example: $\frac{25}{100} = 25\%$

  The ____________________ did not change.

- Example: $\frac{3.8}{100} = 3.8\%$

  The ____________________ is not changed, the decimal point doesn't move if

  the ____________________ is 100.

- To change a decimal to a percent:

1. Move the decimal point to places to the ____________________.
2. Write the ____________________ sign.
   - Example: $0.56 = 56\%$
   - Example: $0.345 = 34.5\%$
   - Example: $0.02 = 2\%$

- To change percents to a decimal number:

1. Move the decimal two places to the ____________________.

2. Delete the ____________________ sign.

    - Example: 97% = 0.97
    - Example: 68.5% = 0.685
    - Example: 0.64% = 0.0064

## Fractions and Percents

- If a fraction has denominator ____________________, it can be changed to a percent by writing the ____________________ and adding the ____________________ sign.
- If the denominator is a factor of ____________________ (2, 4, 5, 10, 20, 25, or 50), the fraction can be changed to an equivalent fraction with denominator of ____________________ and then changed to a percent.
- When fractions do NOT have factors of 100 as denominators you will need to:
- Change the fraction to decimal form by ____________________, either by long division or with a calculator (depending on instructions from instructor).
- Move the decimal two places to the ____________________ and write the ____________________ symbol.

Example: $\frac{3}{4} = 4\overline{)3} = 0.75 = 75\%$

- Helpful calculation tips:
- The numerator goes ____________________ the division symbol and the denominator goes ____________________ of the division symbol.
- If you are using a calculator, you will type the numerator ____________________, then the division symbol, and then the ____________________, followed by enter/equal.
- To change a percent to a fraction or a mixed number:
- Write the percent as a fraction with 100 as the ____________________ and drop the ____________________ symbol.

- ____________________ the fraction, if possible.

Example: $80\% = \frac{80}{100} = \frac{2 \cdot 2 \cdot 2 \cdot 2 \cdot 5}{2 \cdot 2 \cdot 5 \cdot 5} = \frac{4}{5}$

## Solving Percent Problems by Using the Equation: $R \cdot B = A$

- Basic equation for solving percent problems is $rate \cdot base = amount$ or $R \cdot B = A$
- R= ____________________, always written in ____________________ form
- B = ____________________, the number we are finding the ____________________ of
- A = ____________________, part of the ____________________
  - ____________________ means to multiply $(\cdot)$
  - ____________________ means equals $(=)$
- There are ____________________ basic types of percent problems using $R \cdot B = A$
- Type 1: Find the amount given the ____________________ and the percent.

  Example: What is 75% of 600? $A=0.75 \cdot 600$
- Type 2: Find the base given the ____________________ and the amount.

  Example: 15% of what number is 157.5? $0.15 \cdot B=157.5$
- Type 3: Find the percent given the base and the ____________________.

  Example: What percent of 87 is 115? $R \cdot 87=115$
- The operations in these examples can be performed with a ____________________ or by ____________________.
- Solving it either way, the equations should be written so that the ____________________ signs are aligned one under the other.
- Also, writing the equations and the calculated values will help you remember whether you are ____________________ or ____________________.

## Applications: Discount, Sales Tax, Commission, and Percent Increase/Decrease

Problem solving can involve careful ____________________, ____________________, and some original or independent ____________________.

The basic steps for solving word problems are:

1. ____________________ the problem. For example:
   - Read the ____________________.
   - Understand all the ____________________.
   - If it helps, restate the problem in your ____________________ words.
   - Be sure that there is enough ____________________.
2. Devise a ____________________ using one or all of the following:
   - Guess, ____________________, or make a list of possibilities.
   - ____________________ a picture or diagram.
   - Use a ____________________ and form an equation.
3. Carry out the ____________________. For example:
   - Try all the ____________________ you have listed.
   - ____________________ your picture or diagram for insight into the solution.
   - ____________________ any equation that you may have set up.
4. Look back over the results. For example:
   - Can you see an ____________________ way to solve the problem?
   - Does your solution actually ____________________? Does it make ____________________ in terms of the wording of the problem? Is it ____________________?
   - If there is an equation, check your ____________________ in the equation.

- Keep in mind that a calculator is a ____________________ to ____________________, not ____________________, the necessary skills and abilities related to problem solving.
- A discount is a ____________________ in the original selling price (or ____________________ ____________________) of an item.
- The new, reduced price is called the ____________________ price.
- The discount is the ____________________ between the original price and the sale price.
- The rate of discount (or percent of discount) is a ____________________ of the original price.
- Sales tax is a tax charged on the actual ____________________ price of goods sold by retailers.
- The rate of the sales tax (a ____________________ of the actual selling price) varies from state to state (or even city to city in some cases). States and cities use sales tax to pay for public services.
- A ____________________ is a fee paid to an agent or salesperson for a service.
- Commissions are usually a ____________________ of a negotiated contract or a percent of sales.
- In some cases, salespeople earn a ____________________ commission on what they sell.
- In other cases, the salesperson earns a base salary ____________________ a commission on sales above a certain amount.
- Percent of increase (or the percent of decrease) is the percent a certain value ____________________.

- In many cases the increase in value is called ____________________, and the decrease in value is called ____________________.

## Applications: Simple Interest

- Profit is the ____________________ between selling price and cost.
- Profit = ____________________ price − ____________________
- Two Types of Percent of Profit (profit in the numerator in both types):
    - Percent of profit based on ____________________: (Cost is the denominator.)

      $\frac{profit}{\cos t} = \%$ of profit based on cost

    - Percent of profit based on selling ____________________: (Selling price is the denominator.)

      $\frac{profit}{selling\ price} = \%$ of profit based on selling price

- Interest is money paid for the ____________________ of money. The initial amount of money that is invested or borrowed is called the ____________________. The rate is the percent of ____________________ and is stated as an annual (yearly) rate.
- Interest is either ____________________ or ____________________, depending on whether you are the ____________________ or the ____________________. Some loans (called ____________________) are based on ____________________ interest and involve only one payment (including interest and principal). This payment is paid at the end of the term of the loan, usually for a period of one year or less.
- A formula is a general statement (usually an equation) that relates two or more ____________________. The following formula is used to calculate simple interest: $I = P \cdot r \cdot t$.
- The simple interest formula is $I = P \cdot r \cdot t$ where:
    - $I =$ ____________________ (earned or paid)
    - $P =$ principal (the amount ____________________ or ____________________)

- $r$ = rate of interest (stated as an annual rate) in ____________________ number or ____________________ form
- $t$ = time (one year or fraction of a ____________________)
- Note: For calculation purposes, we will use ____________________ days in one year and 30 days in a month. However, many lending institutions now (with the advent of computers) base their calculations on 365 days per year and pay or charge interest on a daily basis.

## U.S. Measurements

Fill in the table based on the information given in the Learn portion of this lesson:

| U.S. Units of Length | |
|---|---|
| 12 inches (in) = 1 __________ (  ) | 3 feet = 1 __________ (  ) |
| 36 inches = 1 __________ | 5280 feet = 1 __________ (  ) |

| U.S. Units of Weight | |
|---|---|
| 16 ounces = 1 __________ (  ) | 2000 pounds = 1 __________ (  ) |

| U.S. Units of Capacity | |
|---|---|
| 8 fluid ounces (  ) = 1 __________ (  ) | 2 pints = 1 __________ (  ) |
| 2 cups = 1 __________ (  ) | 4 quarts = 1 __________ (  ) |
| 1 pint = 16 __________ __________ (  ) | |

| Units of Time | |
|---|---|
| 60 seconds (  ) = 1 __________ (  ) | 24 hours = 1 __________ |
| 60 minutes = 1 __________ (  ) | 7 days = 1 __________ |

- To convert from one unit of measure, you will either:
  - ____________________ to convert to smaller units. (There will be more smaller units.)
  - ____________________ to convert to larger units. (There will be fewer larger units.)

## The Metric System

Fill in the table based on the information given on length in the Learn portion of this lesson:

| Metric Units of Length |
|---|
| 1 millimeter (   ) = 0.001 _____ (   ) |
| 1 centimeter (   ) = 0.01_____ |
| 1 decimeter (   ) = 0.1 _____ |
| 1 meter (m) = 1.0 m ( the _______________ ) |
| 1 dekameter (   ) = 10 _____ |
| 1 hectometer (   ) = 100 _____ |
| 1 kilometer (   ) = 1000 _____ |

Quick Conversion References for length that would be helpful to memorize:

| |
|---|
| 1 m = 1000 _____ |
| 1 m = 100 _____ |
| 1 m = 10 _____ |

- To change to a measure of length that is larger to smaller:
  - one unit smaller, ____________________ by _____
  - two units smaller, ____________________ by _____
  - three units smaller, ____________________ by _____
  - The pattern will continue…

- To change to a measure of length that is smaller to larger:
  - one unit larger, ______________________ by _____
  - two units larger, ______________________ by _____
  - three units larger, ______________________ by _____
  - The pattern will continue…

Using the information given in Learn, use the line below to create a number line that corresponds with the metric units of measurement.

---

- Keep these things in mind when working with the metric system:
  - A _____ is written to the left of the decimal point if there is no whole number part.
  - No commas are used in writing numbers. If a number has more than ______________________ digits (left or right of the decimal point), the digits are grouped in threes from the decimal point with a ______________________ between the groups.
- Mass is the amount of ______________________ in an object. Regardless of where the object is in space, its mass remains the ______________________.
- Weight is the ______________________ of the Earth's gravitational pull on an object. The farther an object is from Earth, the ______________________ the gravitational pull of the Earth.

Fill in the table based on the information given on mass in the Learn portion of this lesson:

| Measures of Mass |
|---|
| 1 milligram (   ) = 0.001 _____ (   ) |
| 1 centigram (   ) = 0.01 _____ |

| |
|---|
| 1 decigram (    ) = 0.1 _____ |
| 1 gram = 1.0 gram |
| 1 dekagram (    ) = 10 _____ |
| 1 hectogram (    ) = 100 _____ |
| 1 kilogram (    ) = 1000 _____ |
| 1 metric ton (    ) = 1 000 000 _____ |

Quick Conversion References for mass that would be helpful to memorize:

| | |
|---|---|
| 1000 mg = 1 _____ | 0.001 g = 1 _____ |
| 1000 g = 1 _____ | 0.001 kg = 1_____ |
| 1000 kg = 1 _____ | 0.001 t = 1_____ |
| 1 000 000 g = 1 _____ | 1 000 000 000 mg = 1_____ |

Fill in the table based on the information given on capacity in the Learn portion of this lesson:

| Measures of liquid volume |
|---|
| 1 milliliter (    ) = 0.001 _____ (    ) |
| 1 liter = 1.0 _____ |
| 1 hectoliter (    ) = 100 _____ |
| 1 kiloliter (    ) = 1000 _____ |

Quick Conversion References for capacity that would be helpful to memorize:

| | |
|---|---|
| 1000 mL = 1 _____ | 1mL = 1 _____ |
| 1000 L = 1 _____ | 1 L = 1 _____ |
| 10 hL = 1 _____ | 1 kL = 1 _____ |

## U.S. to Metric Conversions

The U.S. customary measure is in degrees ____________________ (° ).

The metric measure is in degrees ____________________ (° ).

The formula for converting Fahrenheit to ____________________ is $C = \frac{5(F-32)}{9}$.

The formula for converting Celsius to ____________________ is $F = \frac{9C}{5} + 32$.

| Length Equivalents (rounded values) | |
|---|---|
| U.S. to Metric | Metric to U.S. |
| 1 in = 2.54 _____ | 1 cm = 0.394 _____ |
| 1 ft = 0.305 _____ | 1 m = 3.28 _____ |
| 1 yd = 0.914 _____ | 1 m = 1.09 _____ |
| 1 mi = 1.61 _____ | 1 km = 0.62 _____ |

| Area Equivalents (rounded values) | |
|---|---|
| U.S. to Metric | Metric to U.S. |
| $1 in^2 = 6.45$ _____ | $1 cm^2 = 0.155$ _____ |
| $1 ft^2 = 0.093$ _____ | $1 m^2 = 10.764$ _____ |
| $1 yd^2 = 0.836$ _____ | $1 m^2 = 1.196$ _____ |
| 1 acre = 0.405 _____ | 1 ha = 2.47 _____ |

| Volume Equivalents (rounded values) | |
|---|---|
| U.S. to Metric | Metric to U.S. |
| $1 in^3 = 16.387$ _____ | $1 cm^3 = 0.06$ _____ |
| $1 ft^3 = 0.028$ _____ | $1 m^3 = 35.315$ _____ |
| 1 qt = 0.946 _____ | 1 L = 1.06 _____ |
| 1 gal = 3.785 _____ | 1 L = 0.264 _____ |

| Mass Equivalents (rounded values) | |
|---|---|

| U.S. to Metric | Metric to U.S. |
|---|---|
| 1 oz = 28.35 _____ | 1 g = 0.035 _____ |
| 1 lb = 0.454 _____ | 1 kg = 2.205 _____ |

## Reading Graphs

- The Purposes of Four Types of Graphs
  - Bar Graphs: to emphasize ____________________ amounts.
  - Circle Graphs: to help in understanding ____________________ or parts of a ____________________. Circle graphs are also called ____________________ charts.
  - Line Graphs: to indicate ____________________ or ____________________ over a period of time.
  - Histograms: to indicate data in ____________________ (a ____________________ or interval of numbers).
- A common characteristic of all graphs is that they are intended to ____________________ information about numerical data ____________________ and ____________________.
- All graphs should:
  - Be clearly ____________________.
  - Be easy to ____________________.
  - Have appropriate ____________________.
- Keep these terms in mind when working with histograms:
  - Class: an ____________________ (or range) of numbers that contains data items.
  - Lower class limit: the ____________________ whole number that belongs to a class.
  - Upper class limit: the ____________________ whole number that belongs to a class.
  - Class boundaries: numbers that are halfway between the ____________________ limit of one class and the ____________________ limit of the next class.
  - Class width: the ____________________ between the class boundaries of a class (the width of each bar).
  - Frequency: the ____________________ of data items in a class.

## The Real Number Line and Inequalities

- A number line is a picture of different types of ____________________ and their relationships to each other.
- The graph of a number is the point that ____________________ to the number and the number is called the ____________________ of the point.
- On a horizontal number line, the point one unit to the left of 0 is the ____________________ of 1. It is called negative 1 and is symbolized $-1$.
- The negative sign ($-$) indicates the ____________________ of a number as well as a ____________________ number. It is also used to indicate ____________________.
- The set of numbers consisting of the ____________________ numbers and their ____________________ is called the set of integers: $Z = \{\ldots, -3, -2, -1, 0, 1, 2, 3, \ldots\}$.
- The ____________________ numbers are also called ____________________ integers. Their opposites are called ____________________ integers. Zero is its ____________________ opposite and is neither positive nor negative ($0 = -0$).
- The opposite of a positive integer is a ____________________ integer, and the opposite of a negative integer is a ____________________ integer.
- A rational number is a number that can be written in the form of $\frac{a}{b}$, where a and b are ____________________ and $b \neq$ ____________________. ($\neq$ is read "is not equal to").
- A rational number is a number that can be written in ____________________ form as a ____________________ decimal or as an infinite ____________________ decimal.
- Irrational numbers can be written as infinite ____________________ decimal numbers.

- All rational numbers and irrational numbers are classified as ____________________ numbers ($\mathbb{R}$) and can be written in some decimal form.
- Symbols of inequalities:
- $=$ is ____________________ to
- $\neq$ is ____________________ equal to
- $<$ is ____________________ than
- $>$ is ____________________ than
- $\leq$ is less than or ____________________ to
- $\geq$ is greater than or ____________________ to

## Introduction to Absolute Value

- Absolute value is the ____________________ a number is from 0 on a number line
- Absolute value is symbolized by two ____________________ bars, $|absolute\ value|$.
- The absolute value of a real number is ____________________ negative.
- Examples:

$|84| =$

$|-17| =$

- The solution sets and the corresponding graphs for absolute value inequalities can be ____________________ or ____________________, depending on the nature of the inequality.
- Graph the set of integers indicated by the given inequality on a real number line.

$|x| > 8$

________________________________________________________________

$|y| \leq -7$

________________________________________________________________

## Exponents

- The type of operation used when working with exponents is ____________________ multiplication.
- In the expression, $6^2$, the number that will be multiplied repeatedly is __________.
- In the equation, $2^4 = 16$ __________ is the base, __________ is the exponent, and__________ is the exponential expression.
- Using the exponential expression, $11^4$, circle the equivalent expression below:

  11·11·11·11 or 4·4·4·4·4·4·4·4·4·4·4

- The exponent of 2 is represented by the word ____________________.
- The exponent of 3 is represented by the word ____________________.
- Exponents other than 2 and 3 can be read using the expression, "____________________ ____________________ (fill in the number) ____________________"
- When the exponent 0 is used for any base, except with a base of 0, the value of the expression is defined to be ____________________.

  Example with the base 2: $2^4 = 16$, $2^3 = 8$, $2^2 = 4$, $2^1 = 2$, $2^0 = 1$.

- Order of Operations
- First, simplify within ____________________ ____________________, such as parentheses ( ), brackets [ ], or braces { }. Start with the ____________________ group.
- Second, evaluate any numbers or expressions indicated by ____________________.
- Third, moving from ____________________ to ____________________, perform any ____________________ or ____________________ in the order in which it appears.

- Fourth, moving from ____________________ to ____________________, perform any ____________________ or ____________________ in the order in which it appears.

## Square Roots

- A number is squared when it is ____________________ by itself.
- If an integer is ____________________, the result is called a ____________________ square.

| Squares of Positive Integers from 1 to 20 (Perfect Squares) | | | | |
|---|---|---|---|---|
| $1^2 = 1$ | $5^2 = 25$ | $9^2 = 81$ | $13^2 = 169$ | $17^2 = 289$ |
| $2^2 = 4$ | $6^2 = 36$ | $10^2 = 100$ | $14^2 = 196$ | $18^2 = 324$ |
| $3^2 = 9$ | $7^2 = 49$ | $11^2 = 121$ | $15^2 = 225$ | $19^2 = 361$ |
| $4^2 = 16$ | $8^2 = 64$ | $12^2 = 144$ | $16^2 = 256$ | $20^2 = 400$ |

- Finding the square root is the ____________________ of squaring a number.
- The symbol $\sqrt{\ }$ is called a ____________________ ____________________.
- The number under the radical sign is called the ____________________.

| Squares Roots of Perfect Squares from 1 to 400 | | | | |
|---|---|---|---|---|
| $\sqrt{1} = 1$ | $\sqrt{25} = 5$ | $\sqrt{81} = 9$ | $\sqrt{169} = 13$ | $\sqrt{289} = 17$ |
| $\sqrt{4} = 2$ | $\sqrt{36} = 6$ | $\sqrt{100} = 10$ | $\sqrt{196} = 14$ | $\sqrt{324} = 18$ |
| $\sqrt{9} = 3$ | $\sqrt{49} = 7$ | $\sqrt{121} = 11$ | $\sqrt{225} = 15$ | $\sqrt{361} = 19$ |
| $\sqrt{16} = 4$ | $\sqrt{64} = 8$ | $\sqrt{144} = 12$ | $\sqrt{256} = 16$ | $\sqrt{400} = 20$ |

- The complete expression, such as $\sqrt{64}$ is called a ____________________.

- Many square roots (square roots of numbers other than perfect squares), are ____________________ numbers, and their decimal forms are infinite ____________________ decimals.
- Real numbers are numbers that are either ____________________ or ____________________. That is, every rational number and every irrational number is a ____________________ number.

## Addition with Real Numbers

- The sum of two positive real numbers is ____________________.

- The sum of two negative real numbers is ____________________.

- The sum of a positive real number and a negative real number may be

  ____________________ or ____________________ (or ____________________)

  depending on which number is ____________________ from 0.

- To add two real numbers with like signs:

1. Add their ____________________ ____________________
2. Use the ____________________ sign

- To add two real numbers with unlike signs:

1. ____________________ their absolute values (the smaller from the larger)
2. Use the sign of the number with the ____________________ absolute value

- The ____________________ sign may be omitted when writing ____________________

  numbers, but the ____________________ sign must always be written for

  ____________________ numbers.

- If there is no sign in front of a real number, the real number is understood to be

  ____________________.

- A number is said to be a ____________________ or to ____________________ an

  equation if it gives a true statement when substituted for the variable.

## Subtraction with Real Numbers

- The opposite of a real number is called its ____________________ ____________________.
- The sum of a number and its additive inverse is ____________________.
- Adding positive numbers "____________________ _____" or "____________________" the numbers in a ____________________ direction (moving ____________________ on the number line).
- Adding negative numbers "____________________ _____" or "____________________" the numbers in a ____________________ direction (moving ____________________ on the number line).
- In subtraction, you want to find the "________________________________________" two numbers.
- To subtract, ____________________ the ____________________ of the number being subtracted.
- To find the change in value between two numbers, take the end value and ____________________ the beginning value.
- This process looks like:

  Change in value = (________________ value) − (________________ value)

## Multiplication and Division with Real Numbers

- If $a$ and $b$ are positive real numbers, then:
  - The product of two positive numbers is ____________________: $a \cdot b = +ab$.

    Example: $8(14) = 122$
  - The product of two negative numbers is ____________________: $(-a)(-b) = +ab$.

    Example: $-4(-5) = 20$
  - The product of a positive number and a negative number is ____________________: $a(-b) = -ab$.

    Example: $10(-3) = -30$
  - The product of 0 and any number is _____: $a \cdot 0 = 0$ and $(-a) \cdot 0 = 0$.

    Example: $71(0) = 0$
- If $a$ and $b$ are positive real numbers (where $b \neq 0$), then:
  - The quotient of two positive numbers is ____________________: $\frac{a}{b} = +\frac{a}{b}$

    Example: $\frac{14}{2} = 7$
  - The quotient of two negative numbers is ____________________: $\frac{-a}{-b} = +\frac{a}{b}$

    Example: $\frac{-18}{-6} = 3$
  - The quotient of a positive number and a negative number is ____________________:

    $\frac{-a}{b} = -\frac{a}{b}$ *AND* $\frac{a}{-b} = -\frac{a}{b}$

    Example: $\frac{-144}{12} = -12$

    Example: $\frac{88}{-11} = -8$

- Quick Reference Facts with Multiplication and Division with Real Numbers:
  - If the numbers have the ____________________ sign, both the product and quotient will be ____________________.
  - If the numbers have ____________________ signs, both the product and quotient will be ____________________.

## Order of Operations with Real Numbers

- Order of Operations:
  - Simplify within grouping symbols, such as

    ____________________ ( ), ____________________ [ ], and

    ____________________ { }, working from the innermost grouping

    ____________________.
  - Find any powers indicated by ____________________.
  - Moving from ____________________ to ____________________, perform any

    ____________________ or ____________________ in the order they appear.
  - Moving from ____________________ to ____________________, perform any

    ____________________ or ____________________ in the order they appear.
- Keep in mind that there are other ____________________ symbols like the

  ____________________ ____________________ bars (such as $|3+5|$ ), the

  ____________________ ____________________ (as in $\frac{14+7}{3}$ ), and the

  ____________________ ____________________ symbol (such as $\sqrt{40+9}$ ).
- Even though the mnemonic PEMDAS is helpful, remember that multiplication

  and division are performed as they ____________________, left to right.
- Also, addition and subtraction are performed as they ____________________, left to right.

## Properties of Real Numbers

- Addition is ____________________ while subtraction is _____ ____________________.
- ____________________ Property
  - Of Addition $a+b=b+a$
  - Of Multiplication $ab=ba$
- ____________________ Property
  - Of Addition $(a+b)+c=a+(b+c)$
  - Of Multiplication $a(bc)=(ab)c$
- ____________________ Property
  - Of Addition $a+0=0+a=a$
  - Of Multiplication $a\cdot 1=1\cdot a=a$
- ____________________ Property
  - Of Addition $a+(-a)=0$
  - Of Multiplication $a\cdot\frac{1}{a}=1$ (for $a\neq 0$)
- ____________________ Factor Law
  - $a\cdot 0=0\cdot a=0$
- ____________________ Property of ____________________ over ____________________
  - $a(b+c)=ab+ac$

## Scientific Notation

- In scientific notation, decimal numbers are written as the ____________________ of a number ____________________ than or equal to 1 and less than 10, and an integer ____________________ of 10.

- In scientific notation there is just ____________________ digit to the left of the decimal point.

- The ____________________ tells how many places the decimal point is to be moved and in what direction.

- If the exponent is ____________________, the decimal point is moved to the right.

- Example: $3 \times 10^2$

  Since the exponent is ____________________, you will move the decimal to the right ____________________ places, giving the answer of 300.

- A ____________________ exponent indicates that the decimal point should move to the left.

- Example: $8.7 \times 10^{-3}$

  Since the exponent is ____________________, you will move the decimal to the left ____________________ places, giving the answer of 0.0087.

- If $N$ is a decimal number, then in scientific notation:

  $N = a \times 10^n$ where $1 \leq a < 10$ and $n$ is an ____________________.

## Simplifying and Evaluating Algebraic Expressions

- A single number is called a ____________________.
- Any ____________________ or ____________________ or the indicated product and/or quotient of constants and variables is called a ____________________.
- The number written next to a variable is called the ____________________ (or the ____________________ ____________________) of the variable.
- If no number is written next to a variable, the coefficient is understood to be _____. If a negative sign (−) is next to a variable, the coefficient is understood to be _____.
- ____________________ ____________________ (or similar terms) are terms that are constants or terms that contain the same variables raised to the same powers.
- If there is more than one variable in a term, then the ____________________ of the ____________________ on the variables is the degree of the term.
- An algebraic expression is a combination of ____________________ and ____________________ using any of the ____________________ of addition, subtraction, multiplication, or division, as well as exponents.
- To combine like terms, add (or subtract) the coefficients and keep the ____________________ variable expression.
- To evaluate an algebraic expression:
  - Combine like terms, if possible.
  - Substitute the values given for any variables.
  - Follow the rules for order of operations.

## Solving Linear Equations: $x + b = c$

- An equation is a statement with two ____________________ expressions that are equal, which means both expressions represent the same number.
- If an equation contains a ____________________, any number that gives a true statement when substituted for the variable is called a ____________________ to the equation.
- The solutions to an equation form a ____________________ _____.
- The process of finding the solution set is called ____________________ the equation.
- Steps to solving equations:

  ____________________ any ____________________ terms on each side of the equation.

  Use the ____________________ principle of equality and add the ____________________ of the constant $b$ to both sides. The objective is to ____________________ the variable on one side of the equation (either the left side or the right side) with a coefficient of +1.

  Check your answer by ____________________ it for the variable in the original equation.

Example:

| | |
|---|---|
| $2x + 11 + 3x = 4x - 13$ | Combine like terms. |
| $5x + 11 = 4x - 13$ | Subtract $4x$ to each side. |
| $x + 11 = -13$ | Subtract 11 to each side. |
| $x = -24$ | Check your answer. |
| $2(-24) + 11 + 3(-24) = 4(-24) - 13$ | Substitute -24 for x in the original equation. |
| $-48 + 11 - 72 = -96 - 13$ | Combine terms. |

$-109 = -109$

Since you get a true statement here, you know that -24 is a solution for the equation.

## Solving Linear Equations: *ax* = *c*

- Steps for solving equations in the form of ax=c:
    - Combine any ____________________ terms on each side of the equation.
    - Use the ____________________ (or division) principle of equality and multiply both sides of the equation by the reciprocal of the coefficient of the variable. (Note: This is the same as dividing both sides of the equation by the coefficient.) Thus the coefficient of the variable will become +1 .
    - Check your answer by ____________________ it for the variable in the original equation.

Example:

$-7x = -56$

$\frac{-7x}{-7} = \frac{-56}{-7}$ Divide both sides by -7.

$x = 8$ Check your answer.

$-7(8) = -56$ Substitute 8 for x in the original equation.

$-56 = -56$ Since this is a true statement, the solution for the equation is 8.

## Solving Proportions

- Steps to solve a proportion:
  - Find the cross ____________________ (or cross ____________________).
  - ____________________ both sides of the equation by the coefficient of the variable.
  - ____________________.
- When you are solving a proportion, it will be helpful for you to write each new equation below the previous equation. Keep the = signs ____________________. This will help you organize the steps and avoid simple ____________________.

Example:

$\frac{2}{6} = \frac{5}{x}$ Use cross products.

$2x = 30$ Divide both sides by 2.

$\frac{2x}{2} = \frac{30}{2}$ Simplify.

$x = 15$

- Steps to solve a word problem with a proportion:
  - Identify the unknown quantity and use a ____________________ to represent this quantity.
  - Set up a ____________________ in which the units are compared in the ____________________ order. (Make sure that the units are labeled so they can be seen to be in the right order.)
  - ____________________ the proportion.

Example:

$\frac{25\ miles}{10\ hours} = \frac{x\ miles}{45\ hours}$ Use cross products.

$10x = 1125$ Divide both sides by 10.

$\frac{10x}{10} = \frac{1125}{10}$ Simplify.

$x = 11.25$

## Perimeter

- A line ____________________ consists of two points on a line and all the points between them.
- A polygon is a ____________________ plane figure, with three or more sides, in which each side is a line segment.
- Each point where two sides meet is called a ____________________.
- A closed figure begins and ends at the ____________________ point.
- A triangle is a polygon with ____________________ sides. 
- A rectangle is a polygon with ____________________ sides in which adjacent sides are ____________________ (meet at a 90° angle). 
- A square is a rectangle in which all four sides are the ____________________ length. 
- A trapezoid is a ____________________-sided polygon with one pair of opposite sides that are ____________________. 
- A parallelogram is a four-sided polygon with ____________________ pairs of opposite sides ____________________. 
- The perimeter P of a polygon is the ____________________ of the lengths of its ____________________.
- Commonly used formulas for perimeter are:
- Triangle  $P=a+b+c$
- Square  $P=4s$
- Rectangle 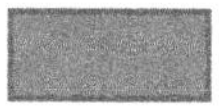 $P=2l+2w$
- Parallelogram  $P=2a+2b$
- Trapezoid 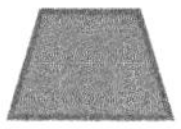 $P=a+b+c+d$

## Area

Area is a measure of the ____________________ of (or surface enclosed by) a plane figure and is measured in ____________________ units.

Units of Area

| Basic U.S. Units of Area | Metric Units of Area |
|---|---|
| Square ____________________ $(in^2)$ | Square ____________________ $(mm^2)$ |
| Square ____________________ $(ft^2)$ | Square ____________________ $(cm^2)$ |
| Square ____________________ $(yd^2)$ | Square ____________________ $(km^2)$ |
| Square ____________________ $(mi^2)$ | Square ____________________ $(m^2)$ |

- Be sure to label your answers with the ____________________ units.
- In triangles and other figures, we have used the letter _____ to represent the height of the figure. The height is also called the ____________________ and is perpendicular to the base.
- Common area formulas:

____________________ $A = \frac{bh}{2}$

____________________ $A = l \cdot w$

____________________ $A = s^2$

____________________ $A = b \cdot h$

____________________ $A = \frac{h(b+c)}{2}$

## Circles

- A ____________________ the set of all points in a plane that are equidistant from a fixed point called the center of the circle.
- The distance from the center of a circle to any point on the circle is known as the ____________________. (The letter r is used to represent the radius of a circle.)
- The ____________________ is the distance from one point on a circle through the center to the point directly opposite it. (The letter d is used to represent the ____________________ of a circle.)
- The perimeter of a circle is called the ____________________.

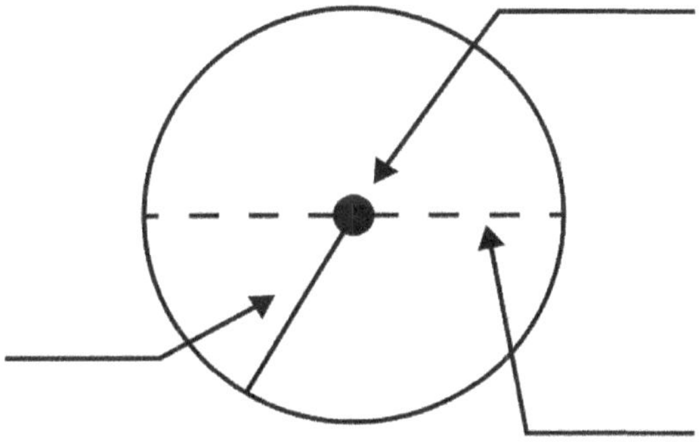

- Formulas for circles are:

____________________ $C = \pi d \text{ or } C = 2\pi r$

____________________ $A = \pi r^2$

- Be aware that 3.14 is only an ____________________ for ____________________ and that related answers are only approximations.
- A semicircle is ____________________ of a circle.

## Volume

- ____________________ is the measure of the space enclosed by a three-dimensional figure and is measured in cubic units.

Units of Volume

| Basic U.S. Units of Volume | Basic Metric Units of Volume |
|---|---|
| Cubic ____________________ ( $in.^3$ ) | Cubic ____________________ ( $mm^3$ ) |
| Cubic ____________________ ( $ft^3$ ) | Cubic ____________________ ( $cm^3$ ) |
| Cubic ____________________ ( $yd^3$ ) | Cubic ____________________ ( $m^3$ ) |

- Commonly Used Volume Formulas

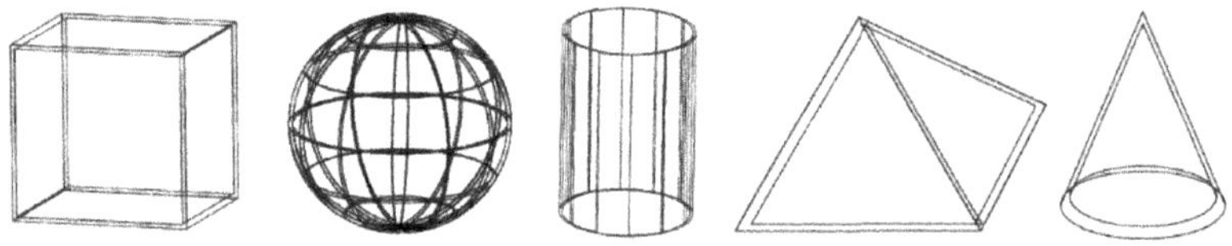

____________________ $v = l \cdot w \cdot h$

____________________ $v = \frac{1}{3}lwh$

____________________ $v = \pi r^2 h$

____________________ $v = \frac{1}{3}\pi r^2 h$

____________________ $v = \frac{4}{3}\pi r^3$

# Translating English Phrases and Algebraic Expressions

| Addition | Subtraction | Multiplication | Division | Exponent (Powers) |
|---|---|---|---|---|
| Add | Subtract (from) | Multiply | Divide | Square of |
| Sum | Difference | Product | Quotient | Cube of |
| Plus | Minus | Times | | |
| More than | Less than | Twice | | |
| Increased by | Decreased by | Of (with fractions and percents) | | |
| | less | | | |

Key Words for Translating Phrases

- Remember that in each case "a number" or "the number" implies the use of a __________ (an unknown quantity).
- Be very careful when writing and/or interpreting expressions indicating________. Be sure that the subtraction is in the order indicated by the wording in the problem. The same is true with expressions involving ________.
- __________________ and ____________________ are done with the values in the same order that they are given in the problem.

## Solving Linear Equations: $x+b=c$

- An _______________ is a statement that two algebraic expressions are equal.
- Any number that gives a true statement when substituted for the variable is called a _____________to the equation.
- The solutions to an equation form a______________ set.
- The process of finding the solution set is called _______________ the equation.
- If $a$, $b$, and $c$ are ________________ and $a \neq 0$, then a linear equation in $x$ is an equation that can be written in the form $ax+b=c$.
- A linear equation in $x$ is also called a _________________- ________________ equation in $x$ because the variable $x$ can be written with the exponent 1. That is, $x = x^1$.
- Addition Principal of Equality
  - If the same algebraic expression is added to ____________ sides of an equation, the new equation has the same solutions as the original equation.
  - If $A$, $B$, and $C$ are algebraic expressions, then the equations, $A=B$ and $A+C=B+C$, have the ____________ solutions.

## Solving Linear Equations: *ax=c*

- The objective is to get the variable by itself on _______________ side of the equation.
- Steps for solving ax=c linear equations:

1. Combine any terms on each side of the equation.

2. Use the _______ (or division) principle of equality and multiply sides of the equation by the reciprocal of the coefficient of the variable. (Note: This is the same as dividing both sides of the equation by the coefficient.) Thus the __________ of the variable will become ___________.

3. Check your answer by _____________ it for the variable in the original equation.

Examples:

| | |
|---|---|
| Solve $18x = -36$ | Divide both sides by 18. |
| $\frac{18x}{18} = \frac{-36}{18}$ | Simplify. |
| $x = -2$ | Check your answer by substituting -2 for x in the original equation. |
| $18(-2) = -36$ | Simplify. |
| $-36 = -36$ | Since this is a true statement, your solution of -2 is correct. |
| Solve $\frac{x}{-2} = -8.4$ | Multiply both sides by -2. |
| $2(\frac{x}{-2}) = -8.4(-2)$ | Simplify. |
| $x = 16.8$ | Check your answer by substituting 16.8 for x in the original equation. |

| | |
|---|---|
| $\frac{16.8}{-2} = -8.4$ | Simplify. |
| $-8.4 = -8.4$ | Since this is a true statement, your solution of 16.8 is correct. |

## Solving Linear Equations: *ax+b=c*

- The general procedure for solving linear equations is now a ________________ of the procedures stated in another section.

- Steps for solving equations in the ax+b=c format.

  1. ______________ like terms on both sides of the equation.

  2. Use the ______________ principle of equality and add the opposite of the constant *b* to both sides.

  3. Use the ______________ (or division) principle of equality to multiply both sides by the reciprocal of the coefficient of the variable (or divide both sides by the coefficient itself). The coefficient of the variable will become +1.

  4. Check your answer by ______________ it for the variable in the original equation.

- Keep in mind that checking can be quite ____________________-__________________ and need not be done for every problem. This is particularly important on _____________. You should check only if you have time after the entire exam is ______________.

Example:

$3x-8=28$ Add 8 to each side.

$3x=36$ Divide both sides by 3.

$x=12$

## Solving Linear Equations: *ax+b=cx+d*

- Remember that the objective is to get the variable on one side of the equation by _________ with a coefficient of +1.
- Steps for Solving Linear Equations that Simplify to the Form ax+b=cx+d

1. Simplify by removing any _________________ symbols and _____________ like terms on each side of the equation.
2. Use the addition principle of equality and add the opposite of a constant term and/or variable term to both sides so that _________________ are on one side and _______ are on the other side.
3. Use the multiplication (or division) principle of equality to multiply both sides by the reciprocal of the coefficient of the variable (or divide both sides by the coefficient itself). The _________________ of the variable will become +1.
4. Check your answer by ______________ it for the variable in the original equation.

| Type of Equation | Number of Solutions | What Does This Mean |
|---|---|---|
| | Finite number of solutions | x=2 |
| | Infinite number of solutions | 5=5 |
| | No Solution | 0=-7 |

- Every linear equation has exactly ______________ solution.
- Every linear equation is a ______________ equation.

# Working with Formulas

- Formulas are general ____________ or principles stated ______________.

Below is a table of formulas and their meanings. You will use these in this lesson.

| Formula | Meaning |
|---|---|
| $I=Prt$ | The simple interest I earned by investing money is equal to the product of the __________P times the rate of interest r times the time t in one year or part of a year. |
| $C=\frac{5}{9}(F-32)$ | Temperature in degrees Celsius C equals $\frac{5}{9}$ times the __________between the Fahrenheit temperature F and __________. |
| $d=rt$ | The __________ traveled d equals the product of the rate of __________ r and the time t. |
| $P=a+b+c$ | The perimeter P of a triangle is equal to the sum of the __________ of the three sides a, b, and c. |
| $F=kAv^2$ | The lifting __________ F is equal to the constant k by the area A and by the square of the plane's velocity v. |

## Applications: Number Problems and Consecutive Integers

- A set of integers consists of the ________________ numbers and their _______________.

  $$Z=\{\ldots,-4,-3,-2,-1,0,1,2,3,4,\ldots\}$$

- Even integers are integers that are divisible by _________________. The even integers are

  $$\{\ldots,-6,-4,-2,0,2,4,6,\ldots\}.$$

- Odd integers are integers that are not __________________. If an odd integer is divided by 2 the remainder will be ______________. The odd integers are

  $$\{\ldots,-5,-3,-1,1,3,5,\ldots\}.$$

- Even integers are consecutive if each is ________________more than the previous ____________ integer. Three consecutive even integers can be represented as $n$, $n+2$, and $n+4$ where $n$ is an even integer.

- Odd integers are consecutive if each is _________________ more than the previous __________________ integer. Three consecutive odd integers can be represented as $n$, $n+2$, and $n+4$ where $n$ is an odd integer.

## Applications: Distance-Rate-Time, Interest, Average

- Word problems are designed to teach you to ________________ carefully, to ___________, and to ________________ clearly.

- Problems involving _____________ usually make use of the relationship indicated by the formula $d = rt$, where $d =$ ______________, $r =$ _____________, and $t =$ ________________.

- When using averages to find a missing test score, let x be represent the ______________ score. You will need to ________________ all of the test scores, including the one that is unknown. Next,________________the sum by the total number of scores, including the one that is unknown. Then, you will solve the equation to find the unknown score.

- The formula $S–C=P$ means _____________ ________________ – ________________= profit.

## Solving Absolute Value Equations

- The absolute value of a number is its ________________ from 0 on a number line.
- Equations involving absolute value may have more than _____________ solution.
- Any number and its opposite have the ________________ absolute value.
- When you solve equations with absolute value for $c > 0$:
  - If $|x| = c$, then _________________ or _______________.
  - If $|ax + b| = c$, then ______________or _______________.
- If the absolute value expression is isolated on one side of the equation, we say that the equation is in ___________ form.

Example:

$|2x+1| = 3$

$2x+1=3$ or $2x+1=-3$

$2x = 2$ $\qquad$ $2x = -4$

$x = 1$ $\qquad$ $x = -2$

# Linear Inequalities

- A ________________ is a collection of objects or numbers.
- The items in the set are called _________________.
- Sets are indicated with _________________ , that look like _________________.
- Elements are listed within the braces the set is said to be in ____________ form.
- The symbol ∈ is read "is an _____________ of" and is used to indicate that a particular number belongs to a _____________.
- If the elements in a set can be __________________ the set is said to be ______________.
- If the elements cannot be counted the set is said to be ________________.
- If a set has absolutely no elements, it is called the ______________ set or null set and is written in the form $\{\ \}$ or with the special symbol ____________.
- The notation ______________ is read "the set of all $x$ such that …" and is called __________________-builder notation.
- The vertical bar that looks like _________________ is read "such that."
- Example: $\{x \mid x\, is\, an\, even\, integer\}$ is read "the set of all $x$ ______________ that $x$ is an even integer."
- The _______________ (symbolized ⋃, as in A ⋃ B) of two (or more) sets is the set of all elements that belong to either ____________ set or the __________set or to ____________ sets.
- The _______________ (symbolized ⋂, as in A ⋂ B) of two (or more) sets is the set of all elements that belong to _________________ sets.

- The word __________ is used to indicate union and the word ___________ is used to indicate intersection.
- Suppose that $a$ and $b$ are two real numbers and that $a<b$ . The set of all real numbers between $a$ and $b$ is called an __________________ of real numbers.
- A ________________ interval such as $-3<x<6$ is read "$x$ is greater than $-3$ and $x$ is less than 6."

Complete the table below based on the information in Learn:

| Type of Interval | Algebraic Notation | Interval Notation | Graph |
|---|---|---|---|
| | | | |
| | | | |
| | | | |
| | | | |
| | | | |

- In an open interval, ______________ endpoint is included.
- In a closed interval, ______________ endpoints are included.
- In a half-open interval, only _____________ endpoint is included.
- The symbol for infinity, which looks like _______________ or _______________,is not a number. It is used to indicate that the interval is to include _________________ real numbers from some point on (either in the positive

direction or the negative direction) without ____________.

- When solving linear inequalities, ________________ or _______________ both sides of an inequality by a negative number causes the "sense" of the inequality to be __________________.
- Rules for solving linear inequalities:
  1. Simplify each side of the inequality by removing any _____________ symbols and _______________ like terms.
  2. Use the ______________ property of equality to add the opposites of constants or variable expressions so that variable expressions are on one side of the inequality and constants are on the other.
  3. Use the ______________ property of equality to multiply both sides by the reciprocal of the coefficient of the variable (that is, divide both sides by the coefficient) so that the new coefficient is 1. If this coefficient is negative, reverse the sense of the inequality.
  4. A quick (and generally satisfactory) check is to select any one number in your solution and _________________ it into the original inequality.
- Compound inequalities have _________________ parts and can arise when a variable or variable expression is to be between two numbers.

## The Cartesian Coordinate System

- The Cartesian coordinate system (or the rectangular coordinate system) is based on a relationship between points in a plane and ordered pairs of real numbers.
- For an equation written like $y = 5x + 7$, ordered pairs are written in the form of ($x$, $y$).
- In an ordered pair, $x$ is called the first coordinate and $y$ is called the second coordinate.
- To find ordered pairs that satisfy an equation in two variables, we can choose any value for one variable and find the corresponding value for the other variable by substituting into the equation.
- There is an Infinite number of such ordered pairs. Any real number could have been chosen for $x$ and the corresponding value for $y$ calculated.
- In an ordered pair of the form $(x,y)$, the first coordinate $x$ is called the Independent variable and the second coordinate $y$ is called the Dependent variable.
- Label the following information on the coordinate plane below: $x$-axis, $y$-axis, quadrants (4), and origin.

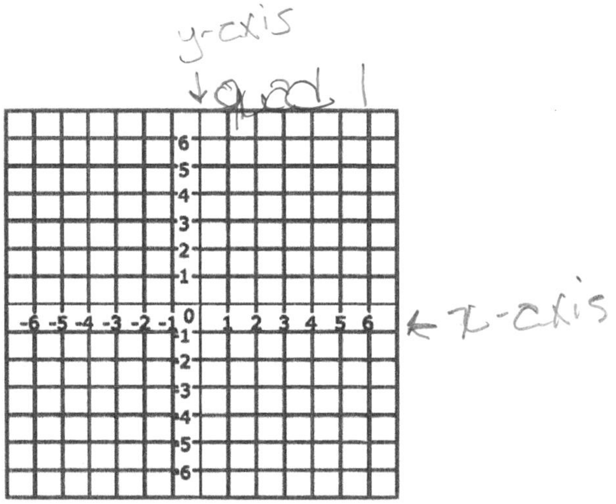


- In quadrant I, the $x$ is positive and the $y$ is positive.

- In quadrant II, the $x$ is Negative and the $y$ is Positive.
- In quadrant III, the $x$ is Negative and the $y$ is Negative.
- In quadrant IV, the $x$ is Positive and the $y$ is Negative.
- There is a one-to-one Correspondence between points in a plane and ordered pairs of real numbers.
- Unless otherwise stated, assume that the grid lines are One unit apart.
- To plot an $x$-coordinate, you will move left for negative values and Right for positive values.
- To plot any $y$-coordinate, you will move down for negative values and up for up values.

## Graphing Linear Equations in Two Variables

- To find some solutions for an equation like $2 - 2x = y$, you will need to:

    1. choose arbitrary values for x
    2. find the corresponding values for y by substituting into the equation.

- The standard form of an equation is Ax+By=C where $A$, $B$, and $C$ are real numbers and $A$ and $B$ are not both equal to zero, is called the standard form of a linear equation.

- In the standard form Ax+By=C, $A$ and $B$ may be positive, negative, or 0, but $A$ and $B$ cannot both equal 0.

- Every line corresponds to some Linear equation, and the graph of every linear equation is a line.

- Two points determine a line.

- Steps to graph a linear equation in two variables:

    1. Locate any two points that satisfy the equation. (Choose values for $x$ and $y$ that lead to simple solutions. Remember that there is an infinite number of choices for either $x$ or $y$. But, once a value for $x$ or $y$ is chosen, the corresponding value for the other variable is found by substituting into the equation.)
    2. Plot these two points on a Cartesian coordinate system.
    3. Draw a line through these two points. (Note: Every point on that line will satisfy the equation.)

4. To check: Locate a ___third___ point that satisfies the equation and check to see that it does indeed lie on the line.

- By letting $x$ = ___0___, you will locate the point on the graph where the line crosses (or intercepts) the $y$-axis. This point is called the $y$-___intercept___ and is of the form (_0_, $y$).
- The $x$-intercept is the point found by letting $y$ = ___0___ .This is the point where the line crosses (or intercepts) the $x$-axis and is of the form ($x$, _0_).

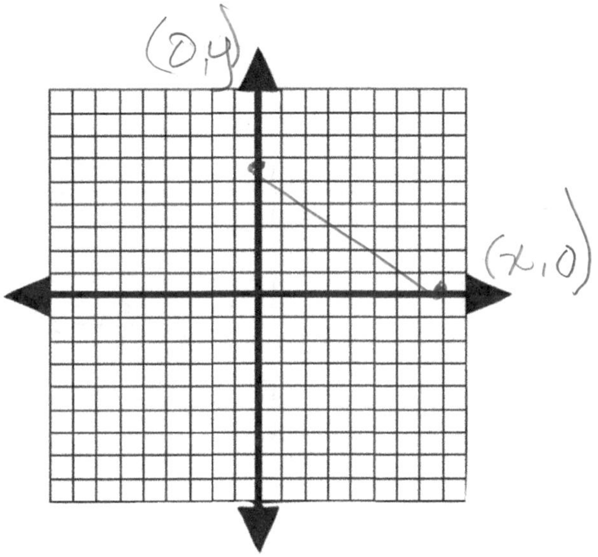


- When the intercepts result in a point with fractional (or decimal) coordinates and ___estimation___ is involved, then a ___third___ point that satisfies the equation should be found to verify that the line is graphed correctly.
- For real numbers $a$ and $b$, the graph of $y = b$ is a ___Horizontal___ line and $x = a$ is a ___Vertical___ line.

## The Slope-Intercept Form

- For a line, the ratio of __rise__ to __run__ is called the slope of the line.

- Plot the rise and run between points on the line in the graph below:

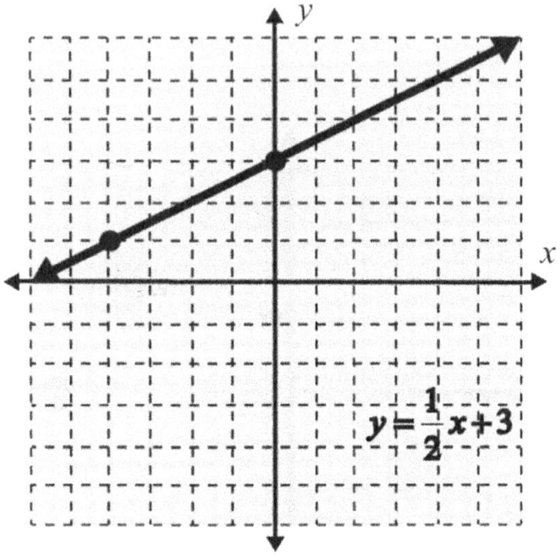


$\frac{\text{RISE}}{\text{RUN}} = \frac{2}{4} = \frac{1}{2}$

- The concept of slope also relates to situations that involve __rate__ of __change__.

- __Slope__ Formula $= m = \dfrac{rise}{run} = \dfrac{y_2 - y_1}{x_2 - x_1}$

- The variable that represents slope is __m__.

- The slope is the same even if the order of the points is __reversed__.

- The coordinates must be __subtracted__ in the same order in both the __numerator__ and the __denominator__.

- Lines with positive slope go up (__increase__) as we move along the line from __left__ to __right__.

- Lines with negative slope go down(<u>decrease</u>) as we move along the line from <u>left</u> to <u>right</u>.
- If two points have the same <u>y coord</u>, such as (2, 8) and (7,8), then the line through these two points will be <u>horizontal</u>.
- If two points have the same <u>x coord</u>, such as (2, 3) and (2,−2), then the line through these two points will be <u>vertical</u>.
- The following two general statements are true for horizontal and vertical lines:

  1. For horizontal lines (of the form $y = b$), the slope is <u>0</u>.
  2. For vertical lines (of the form $x = a$), the slope is <u>undefined</u>.

- $y=mx+b$ is called the <u>Slope</u>-<u>Intercept</u> form for the equation of a line, where $m$ is the <u>Slope</u> and $(0, b)$ is the <u>y intercept</u>.

## Finding the Equation of a Line

- The standard form for a linear equation is Ax+By=C.
- The slope-intercept form for a linear equation is y=mx+b.
- The point- slope form for a linear equation is $y - y_1 = m(x - x_1)$
- Use the point-slope form when given the SLOPE and an ORDERED pair for a line.
- You will SUBSTITUTE the values for slope and the ordered pair into the equation and then solve.
- To find the equation of a line given two points on the line:

1. Use the formula $m = \frac{y_2 - y_1}{x_2 - x_1}$ to find the SLOPE of the two points.
2. Use this slope, m, and either POINT in the point-slope formula $y - y_1 = m(x - x_1)$ to find the equation.

## The Point-Slope Form

- Parallel lines are lines that never intersect (never cross each other) and these lines have the same slope.
- All vertical lines (undefined slopes) are parallel to one another.
- Perpendicular lines are lines that intersect at 90 degrees (right) angles and whose slopes are negative reciprocals of each other.
- Horizontal lines are perpendiculars to vertical lines.
- To find the equation of the line through a specific point and parallel to a specific line you will need to:

  1. Isolate m in the linear equation given.
  2. Substitute the slope and one ordered pair for a point in the point-slope form.
  3. Solve the equation.

- To find the equation of the line through a specific point and perpendicular to a specific line you will need to:

  1. Isolate m in the linear equation given.
  2. Substitute the opposite reciprocal of the slope and substitute one ordered pair for a point in the point-slope form.
  3. Solve the equation.

- Please complete the table below as a summary of formulas and properties of straight lines:

| | |
|---|---|
| $Ax+By=C$ | Standard Form |
| $m=\frac{y2-y1}{x2-x1}$ | Slope of a Line |
| $y=mx+b$ | Slope-Intercept Form |
| $y-y1\ m=(x-x1)$ | Point-Slope Form |
| $y=b$ | Horizontal Line, slope = 0 |
| $x=a$ | Vertical Line, slope is undefined |
| parallel lines | Have the same slope |
| perpendicular lines | Have slopes that are negative reciprocals of each other |

## Graphing Linear Inequalities in Two Variables

- A straight line separates a plane into __2__ half-planes.

  The points on one side of the line are in one of the __half planes__, and the points on the other side of the line are in the other half-plane.

- The line itself is called the __boundary__ __line__.

- If the boundary line is included in the solution set, then the half-plane is said to be __closed__ (__solid__ line).

- If the boundary line is not included in the solution set, then the half-plane is said to be __open__ (__dashed__ line).

- Steps to Graphing Linear Inequalities:

  1. First, graph the __boundary__ line (__dashed__ if the inequality is < or >, __solid__ if the inequality is ≥ or ≤).
  2. Next, determine which side of the line to __shade__ using one of the following methods.

     **Method 1**

     a. Test any one __point__ obviously on one side of the line.

     b. If the test-point satisfies the inequality, shade the half-plane on that side of the line. Otherwise, shade the other half-plane.

Note: The point (0,0), if it is not on the boundary line, is usually the __easiest__ point to test.

**Method 2**

a. Solve the inequality for ___y___ (assuming that the line is not vertical).

b. If the solution shows $y<$ or $y\leq$, then shade the half-plane ___below___ the line.

c. If the solution shows $y>$ or $y\geq$, then shade the half-plane ___above___ the line.

Note: If the boundary line is ___vertical___, then solve for x. If the solution shows $x>$ or $x\geq$, then shade the half-plane to the ___right___. If the solution shows $x<$ or $x\leq$, then shade the half-plane to the ___left___.

3. The shaded half-plane (and the line if it is solid) is the ___solution___ to the inequality.

## Introduction to Functions and Function Notation

- In Function notation, instead of writing y we write $f(x)$, which is read "f of x."

- To evaluate the equation $f(x) = 5x - 4$ at $f(2)$, we replace x with 2 in the expression $5x - 4$ and simplify. This gives $f(2) = 5(2) - 4 = 6$.

- Using function notation, the ordered pair (2,6) can be written as $(x, f(x))$.

- Given the graph of $y = f(x)$, we can find $f(x)$ for a value of $x$ by determining the y-coordinate of the point on the graph with the given $x$-coordinate.

## Systems of Linear Equations: Solutions by Graphing

- Graphing systems of two equations in two variables is helpful in visualizing the relationship between the equations.
- A solution of a system of linear equations is an ordered pair (or point) that satisfies both equations.
- There are three methods for solving a system of linear equations.

  1. graphing
  2. substitution
  3. Addition

- System of linear equations will be classified as one of the following:

| | |
|---|---|
| Consistent | Has exactly one solution |
| Inconsistent | Has no solution |
| Dependent | Has an infinite number of solutions |

- Steps to solving a system of linear equations by graphing:

  1. Graph both linear equations on the same set of axes.
  2. Observe the point of intersection (if there is one).
     a. If the slopes of the two lines are different, then the lines intersect in one and only one point. The system has a single point as its solution.
     b. If the lines have the same slope and different $y$-intercepts, then the lines are parallel. The system has no solution.

c. If the lines are the ___same___ line, then all the points on the line constitute the solution. There are an ___infinite___ number of solutions.

3. Check the solution (if there is one) in both of the original equations.

## Systems of Linear Equations: Solutions by Substitution

- The objective in the substitution method is to eliminate one of the variables so that a new equation is formed with just one variable.
- If this new equation has one solution then the solution to the system is a single point.
- If this new equation is never true, then the system has no solution.
- If this new equation is always true, then the system has an infinite number of solutions.
- Steps to solving a system of linear equations by substitution:

  1. Solve one of the equations for one of the variables.
  2. Substitute the resulting expression into the other equation.
  3. Solve this new equation, if possible, and then substitute back into one of the original equations to find the value of the other variable. (This is known as back substitution.)
  4. Check the solution in both of the original equations.

## Systems of Linear Equations: Solutions by Addition

- Solving a system of linear equations by addition is also a method of elimination.
- In the method of addition, as with the method of substitution, the objective is to eliminate one of the variables so that a new equation is found with just one variable, if possible.
- If this new equation:
  - has one solution, the solution to the system is a single point.
  - is never true, the system has no solution.
  - is always true, the system has infinite solutions.
- Steps to solve a system of linear equations by addition are:
  1. Write the equations in vertical form, one above the other, so that like terms are aligned.
  2. Multiply all terms of one equation by a constant (and possibly all terms of the other equation by another constant) so that two like terms have opposite coefficients.
  3. Add the two equations by combining like terms and solve the resulting equation, if possible.
  4. Back substitute into one of the original equations to find the value of the other variable.
  5. Check the solution (if there is one) in both of the original equations.
- Guidelines for Deciding Which Method to Use When Solving a System of Linear Equations:

- The graphing method is helpful in "seeing" the geometric relationship between the lines and finding approximate solutions. A calculator can be very helpful here.
- Both the substitution method and the addition method give exact solutions.
- The substitution method may be reasonable and efficient if one of the coefficients of one of the variables is 1.
- The addition method is particularly efficient if the coefficients for one of the variables are opposites.

## Applications: Distance-Rate-Time, Number Problems, Amounts, and Costs

- For applications you will need to create your own systems of equations.
- You should consider making tables when working with applications. The tables can help you collect the information in a more understandable form.
- Basic table set-up will look similar to one of these:

| | Rate | ×Time | =Distance |
|---|---|---|---|
| Vehicle 1 | speed | how long | how far |
| Vehicle 2 | speed | how long | how far |

| Items | Number of Each Item | Value | Total Value |
|---|---|---|---|
| Item A | N | 0.05 | 0.05N |
| Item B | Q | 0.25 | 0.25Q |

## Applications: Interest and Mixture

- Complete the table below of things to remember when working interest problems:

| | |
|---|---|
| P | principal |
| r | rate |
| t | time |
| I | Interest |

- When working on solution problems, your table will be set-up similar to the one below:

| Types of Solutions | Amount of solution | Times × | Percent of a specific item | Equals = | Amount of a specific item |
|---|---|---|---|---|---|
| Type A | | | | | |
| Type B | | | | | |

## Exponents

- If $a$ is a ________________ real number and $m$ and $n$ are integers, then $a^m + a^n = a^{m+n}$.
- What this means is that in order to multiply two powers with the same base, keep the ________________ and ______________ the exponents.

  Example: $2^3 \cdot 2^5 = 2^{3+5} = 2^8$

- If a variable or constant has no exponent written, the exponent is understood to be ________________.
- Any non-zero value raised to the 0 power is worth ____________.
- The expression $0^0$ is ____________.
- To divide two powers with the same base, keep the ____________ and ______________ the exponents.
- In other words, subtract the ______________ exponent from the ______________exponent.

  Example:

$$\frac{d^8}{d^3} = d^{8-3} = d^5$$

- If $a$ is a nonzero real number and $n$ is an ______________, then $a^{-n} = \frac{1}{a^n}$.

  Example:

$$n^{-2} = \frac{1}{n^2}$$

## Exponents and Scientific Notation

- If a is a nonzero real number and m and n are ________________, then $\left(a^m\right)^n = a^{m(n)}$
- In other words, the value of a power raised to a power can be found by ______________ the exponents and ___________ the base.

Example: $(x^2)^5 = x^{2(5)} = x^{10}$

- If a and b are nonzero ___________ numbers and n is an integer then $(ab)^n = a^n b^n$.
- In words, a power of a product is found by raising each factor to that ________________.

Example: $(xy)^3 = x^3 y^3$

- If a and b are nonzero real numbers and n is an integer, then $\left(\frac{a}{b}\right)^n = \frac{a^n}{b^n}$.

Complete the table below as a summary of Exponent Rules:

| | |
|---|---|
| The exponent of 1 | |
| The exponent of 0 | |
| The Product Rule | |
| The Quotient Rule | |
| Negative Exponents | |
| Power Rule | |
| Power of a Product | |
| Power of a Quotient | |

- In words, a power of a quotient (in fraction form) is found by raising both the

  ________________ and the ________________ to that power.

  Example: $\left(\frac{2}{7}\right)^3 = \frac{2^3}{7^3}$

- When using ____________ ________________ and N is a decimal number, then

  in scientific notation $N = a \times 10^n$ where $1 \leq a < 10$ and n is an integer.

- In scientific notation:

  - A ______________ exponent indicates that the decimal point should move to the left.

  - A ______________ exponent indicates that the decimal point should move to the right.

## Introduction to Polynomials

- A term is an expression that involves only _____________ and/or _________________ with constants and/or variables.
- A number written next to a variable indicates ________________, and the number is called the _______________ coefficient (or coefficient) of the variable.
- A ________________ in x is a term of the form $kx^n$ where k is a real number and n is a whole number.
- In the term $kx^n$, n is called the ________________ of the term, and k is called the coefficient.
- A monomial may have more than one variable, and the degree of such a monomial is the ____________ of the degrees of its variables.
- A polynomial is a monomial or the indicated sum or difference of ____________.
- The degree of a polynomial is the ______________ of the degrees of its terms.
- The coefficient of the term of the largest degree is called the _________________ coefficient.
- Use the information in Learn to complete the table below:

| Term | Definition | Example |
|---|---|---|
| | | |

| | | |
|---|---|---|
| Monomial | | |
| Binomial | | |
| Trinomial | | |

- When working with polynomials, terms are written in ________________ order.
- To evaluate a polynomial for a given value of the variable, substitute the value for the ____________ wherever it occurs in the polynomial and follow the rules for order of operations.

## Addition and Subtraction of Polynomials

- The sum of two or more polynomials is found by ______________ like terms.
- Like terms have the same variable raised to the _______________ power.
    - Examples of ____________ terms: $-7m^3, m^3, 16m^3$
    - Examples of _____________ terms: $12x, -5n^3, p^5$
- When adding and subtracting polynomials, the polynomials may be written

  _________________ or ________________.

  Example of writing the problem ________________:

  $(5x^4+3x^2-4x+2)+(6x^4-x^3-7x^2+9x+1)$

  $(5x^4+6x^4)-x^3+(3x^2-7x^2)+(-4x+9x)+(2+1)$

  $11x^4-x^3-4x^2+5x+3$

  Example of writing the problem ________________:

  $(3y^3+2y-4)-(2y^3-7y^2+3y-5)$

  $$\begin{array}{rr} & 3y^3+0y^2+2y-4 \\ + & -2y^3+7y^2-3y+5 \\ \hline & y^3+7y^2-y+1 \end{array}$$

- Helpful tips on adding and subtracting polynomials:
    - If there is a missing term in one of the polynomials, you can add a term with the coefficient of ______ to hold the place.

- A negative sign written in front of a polynomial in parentheses indicates the

  __________________ of the _____________ polynomial.

## Multiplication with Polynomials

- When multiplying a polynomial by a monomial, use the _______________ property:

  $a(b+c) = ab + ac$

  Example:

  $5x(x+2)$

  $5x(x)+5x(2)$

  $5x^2 + 10x$

- When multiplying two polynomials, the ________________ property is applied by multiplying each term of one polynomial by each term of the other.

- In other words, multiply the _____________ term in the first polynomial to every term in the second polynomial. Then you will multiply the ____________ term in the first polynomial to all of the terms in the second polynomial.

  Example:

  $(5x-7)(x+2)$

  $5x(x)+5x(2)+(-7)(x)+(-7)(2)$

  $5x^2 + 10x - 7x - 14$

## The FOIL Method

1. FOIL is a mnemonic device for help remembering which terms to ___________________ together when working with binomials.

2. What terms do the letters in FOIL represent?

   - F: ______________________
   - O: ______________________
   - I: ______________________
   - L: ______________________

3. Using the binomial, $(x+1)(x+4)$, draw arrows from one term to another to show the terms that will be multiplied together when using the FOIL method.

$$(x+1)(x+4)$$

4. What four pairs of terms will be multiplied if you use the FOIL method to find the product of the binomial, $(b+7)(2b+3)$?

   - ______________________
   - ______________________
   - ______________________
   - ______________________

5. Using the FOIL method, find the product of the following binomials:

$(x+1)(x+4)$

$(b+7)(2b+3)$

## Special Products of Binomials

- When two binomials are in the form of the sum and difference of the same two terms, the product will always be the _______________ of the _____________ of the terms.
- The rule for the difference of squares is ___________________________________________.

Example:

$(x+4)(x-4)$

$x^2-16$

- The squares of integers are called _______________ ___________________.
- The perfect squares that you are recommended to memorize are:

_____________________________________________________________________________

- When the square of a binomial is when the two binomials that are being multiplied are the same, you will:

1. Use the _______________ property
2. Notice that you are __________________ the first term, _____________ the middle term, and __________________ the second term

Example:

$(z-9)(z-9)$

$z^2+18z+81$

- Please avoid the error of not having a middle term by using the ________________ property!

# Division by a Monomial

- Rational expressions are ____________________ in which the ____________________ and ________________________ are polynomials.
- A denominator cannot be the value of ___________________ in a rational expression.
- Circle all of the examples of rational expressions below:

$$\frac{x^3}{2x} \qquad \frac{2}{3} \qquad \frac{15x^2+21}{3x} \qquad \frac{10}{0}$$

- The sum of fractions with the same denominator can be written as a fraction by ____________________ the ______________________ and using the common ____________________.
- Complete the example below by combining the fractions together, using the common denominator:

$$\frac{17}{y}+\frac{3x}{y}-\frac{4z}{y} = \underline{\hspace{6cm}}$$

- If you are asked to divide the rational expression that already has a monomial as a denominator, you will divide each ____________________ in the numerator by the ____________________ ____________________ and then simplify each fraction.
- Using the statement above, show the steps to divide each polynomial by the monomial denominator:

$$\frac{12x^3+6x^2-8x}{4x} = \underline{\hspace{1.5cm}} + \underline{\hspace{1.5cm}} - \underline{\hspace{1.5cm}} =$$

## Greatest Common Factor of Two or More Terms

- The result of multiplication is called the ________________.
- The numbers or expressions being multiplied are called ________________ of the product.
- The reverse of multiplication is called __________________.
- The ____________________________________________________ (GCF) of two or more

  integers is the largest integer that is a factor (or divisor) of _____________ of the integers.
- Steps for Finding the GCF for a set of Terms are:

  1. Find the _______________ ___________________ of all integers and integer

     coefficients.

  2. List all the factors that are common to _______________ terms, including variables.

  3. Choose the ___________________ power of each factor that is common to all terms.

  4. ________________ these powers to find the GCF.
- If there is no common prime factor or variable, then the GCF is ________________.

# Greatest Common Factor of a Polynomial

To find a monomial that is the greatest common factor (GCF) of a polynomial:

1. Find the variable(s) of ______________ degree and the largest integer coefficient that are factors of each term of the polynomial. (The ___________ of these is the greatest common monomial factor.)

2. ______________ this monomial factor into each term of the polynomial, resulting in another polynomial factor.

Example: Factor the polynomial $3x^4 + 18x^3 - 12x^2 + 6x$ by factoring out the GCF.

$3x^4 : 3 \cdot x \cdot x \cdot x \cdot x$

$18x^3 : 2 \cdot 3 \cdot 3 \cdot x \cdot x \cdot x$

$-12x^2 : 2 \cdot 2 \cdot 3 \cdot x \cdot x$

$6x : 2 \cdot 3 \cdot x$

The GCF is $3x$.

$3x(x^3 + 6x^2 - 4x + 2)$

## Factoring Expressions by Grouping

- We can find the binomial factors by ____________.
- This means looking for ________________ factors in each group and then looking for common binomial factors.

Example:

$7x(2x-1)-3(2x-1)$ Notice both terms have ________________ as a factor.

$(7x-3)(2x-1)$ Group the 7x and the -3 together as one ________________.

Example:

$xy+5x+3y+15$

$(xy+5x)+(3y+15)$ Group the terms into ____________ binomials.

$x(y+5)+3(y+5)$ Factor out the ____________ for each binomial.

$(y+5)(x+3)$ _________________ the GCF's into one binomial and the other terms into another.

## Factoring Trinomials with Leading Coefficient 1

- To factor $x^2 + bx + c$, if possible, find an integer pair of ________________ of $c$ whose ____________ is $b$.

  1. If $c$ is ________________, then both factors must have the ________________ sign.
     - Both will be positive if $b$ is ______________.
     - Both will be negative if $b$ is ______________.
  2. If $c$ is ________________, then one factor must be ________________ and the other __________.

- One additional thing to keep in mind is when factoring polynomials, always look for a common ________________ factor first.
- Then, if there is one, remember to include this ________________ monomial factor as part of the answer.
- If there is a common monomial factor, factor out this common monomial factor ___________________ and then factor the remaining ___________________, if possible.
- Not all polynomials are _________________.
- For example, no matter what combinations are tried, $x^2 + 3x + 4$ does not have two binomial factors with integer coefficients. (There are no factors of +4 that will add to +3.) The polynomial is not factorable (or _________________).
- A polynomial is not ______________ if it cannot be factored as the product of polynomials with integer coefficients.

## Factoring Trinomials by the *ac*-Method

- The ac-method of factoring is in reference to the coefficients a and c in the general form $ax^2+bx+c$ and involves the method of factoring by ____________.

- Complete the table below describing each step of the ac-method and an example:

| General Method<br>$ax^2+bx+c$ | Example<br>$2x^2+9x+10$ |
|---|---|
| Step 1: Multiply $a{\cdot}c$. | |
| Step 2: Find two integers whose product is *ac* and whose sum is *b*. If this is not possible, then the trinomial is not factorable. | Find two integers whose product is 20 and whose sum is 9. |
| Step 3: Rewrite the middle term (*bx*) using the two numbers found in Step 2 as coefficients. | Rewrite the middle term (+9x) using +4 and +5 as coefficients. |
| Step 4: Factor by grouping the first two terms and the last two terms. | Factor by grouping the first two terms and the last two terms. |
| Step 5: Factor out the common binomial factor. This will give two binomial factors of the trinomial $ax^2+bx+c$. | Factor out the common binomial factor (x+2). |

- Helpful hints for factoring:
  - When factoring polynomials, always look for the ______________ ________________ ________________ first. Then, if there is one, remember to include this common factor as part of the ____________.
  - To factor completely means to find factors of the polynomial such that none of the factors are themselves ________________.

    - Not all polynomials are _________________. Any polynomial that cannot be factored as the product of polynomials with integer coefficients is not factorable.
    - Factoring can be checked by _____________ the factors. The product should be the original expression.
- No matter which method you use (the ac-method or the trial-and-error method), factoring trinomials takes _________________. With practice you will become more efficient with either method. Make sure to be _________________ and observant.

## Factoring Trinomials by Trial and Error

- There are two methods of factoring trinomials of the form $ax^2+bx+c$ in which the coefficients are restricted to integers:
  - trial-and-error method (a reverse of the ________________ method of multiplication)
  - the ac-method (a form of ________________)

    Example:

    Factor $x^2+6x+5$ by trial and error.

    Notice the last term is ____________ and middle term is ________. You will have to find a set of factors for 5 that will give you a sum of 6.

    Using the factors of ________ and_________, make two binomials: $(x+1)(x+5)$

- To factor completely means to find factors of the polynomial, none of which are themselves factorable.
  - An example of this would be $2x^2+12x+10=(2x+10)(x+1)$. This is ____________ factored completely because $2x+10=2(x+5)$.
  - We could write $2x^2+12x+10=(2x+10)(x+1)=2(x+5)(x+1)$.
  - This problem can be avoided by first factoring out the GCF (in this case, _________________________).

Example: Factor $2x^2+18x+36$ by trial and error.

Notice there is a common factor of ___________ that needs to be factored out before beginning the trial and error method.

Your new trinomial to factor is $2(x^2 + 9x + 18)$. Notice the last term is an

________ and the middle term is a __________. The factor pairs for 18 are 1 and 18, 2 and 9, and then 3 and 6. Find a pair of factors for 18 that will give you the sum of 9, which is 3 and 6.

Make your new binomials using the 3 and 3: $2(x+3)(x+6)$.

Be sure not to forget to write down the initial GCF of ________ that you factored out at the very beginning of this process.

## Special Factorizations - Squares

- Remember that the _______________ of squares rule is $x^2 - a^2 = (x+a)(x-a)$.
- When factoring difference of squares:

    1. Take the square root of the _______________ term.

    2. Take the square root of the _______________ term.

    3. Form the two binomials, one with an _____________ symbol and one with a _____________ symbol.

- The sum of two squares is an expression of the form $x^2 + a^2$ and is _____________ factorable,

  unless there is a GCF that can be factored out of ________________ terms.

- Factoring a perfect square trinomial gives the __________________ of a binomial.
- In a perfect square trinomial, both the __________________ and _______________ terms of the trinomial must be perfect squares.
- If the first term is of the form $x^2$ and the last term is of the form $a^2$, then the middle term must be of the form ______________ or ________________.

## Special Factorizations - Cubes

- Using the distributive property, show the three steps to arriving at the sum of cubes for $(x+a)(x^2-ax+a^2)$.
  - Step One: ______________________________
  - Step Two: ____________________________
  - Step Three: ____________________________(result: sum of two cubes)
- Using the distributive property, show the three steps to arriving at the difference of cubes for $(x-a)(x^2+ax+a^2)$.
  - Step One: ______________________________
  - Step Two: ____________________________
  - Step Three: ____________________________(result: difference of two cubes)
- Things to remember about sums and difference of cubes:
  - In each case, after multiplying the factors the middle terms drop out and only ___________________ terms are left.
  - The trinomials in parentheses are not perfect square trinomials. These trinomials are not ________________.
  - The sign in the binomial agrees with the sign in the _____________.

## Additional Factoring Practice

- Always look for a common ________________ factor first. If the leading coefficient is ____________, factor out a negative monomial even if it is just −1.
- Check the number of terms.
  - Two terms:
    - __________________ of two squares?—factorable
    - __________________ of two squares?—not factorable
    - __________________ of two cubes?—factorable
    - __________________ of two cubes?—factorable
  - Three terms:
    - perfect square trinomial?
    - use trial-and-error method?
      - Guidelines for the trial-and-error method
      - If the sign of the constant term is positive, the signs in both factors will be the _____________, either both positive or both negative.
      - If the sign of the constant term is negative, the signs in the factors will be ________________, one positive and one negative.
    - use ac-method?
      - Guidelines for the ac-method
        - Multiply ___________ · ___________
        - Find two integers whose ________________ is ac and whose __________________ is b. If this is not possible, the trinomial

is ________________ factorable.

- Rewrite the middle term (bx) using the two numbers found in Step b. as ________________.
- Factor by ________________.

- Four terms: Group terms with a common factor and factor out any ____________ binomial factor.

- Check the possibility of factoring any of the factors.

- Checking: Factoring can be checked by _____________ the factors. The product should be the original expression.

## Solving Quadratic Equations by Factoring

- ________________ equations are equations that can be written in the form $ax^2 + bx + c = 0$ where $a$, $b$, and $c$ are real numbers and $a \neq 0$.
- The form $ax^2 + bx + c = 0$ is called the ________________ form (or general form) of a quadratic equation.
- When solving quadratic equations by factoring, having ________________ on one side before we factor is necessary because of the following zero-factor property.
- In general, a quadratic equation has ____________ solutions.
- In the special cases where the two factors are the same, there is only one solution and it is called a double ________________ (or double root).
- Steps to solving a quadratic equation by factoring are:

    1. ________________ or ______________ terms as necessary so that 0 is on one side of the equation and the equation is in the standard form $ax^2 + bx + c = 0$ where $a$, $b$, and $c$ are real numbers and $a \neq 0$.

    2. ________________ completely. (If there are any fractional coefficients, multiply each term by the least common denominator first so that all coefficients will be integers.)

    3. Set each nonconstant factor equal to ____________ and solve each linear equation for the unknown.

    4. Check each ______________, one at a time, in the original equation.

- Steps to finding an equation that has certain given solutions (or roots):

  1. Rewrite the linear equations with ____________ on one side;

  2. Form the ______________ of the factors and set this product equal to 0.

  3. ________________ the factors. The resulting quadratic equation must have the two given roots (because we know the factors).

## Applications of Quadratic Equations

- Things to consider when solving an application problem:
  - Read the problem carefully at least _______________.
  - Decide what is asked for and assign a ___________________ or _________________ ___________________ to the unknown quantities.
  - ___________________ a chart, table, or diagram relating all the information provided.
  - Form an _________________. (A formula of some type may be necessary.)
  - ___________________ the equation.
  - Check your solution with the wording of the problem to be sure it makes sense.
- Integers are consecutive if each is _______________ more than the previous integer. Three consecutive integers can be represented as ________,____________, and ________.
- Even integers are consecutive if each is 2 more than the previous even integer. Three consecutive even integers can be represented as __________,____________, and ________, where $n$ is an even integer.

  Examples: ___________________________________
- Odd integers are consecutive if each is 2 more than the previous odd integer. Three consecutive odd integers can be represented as __________,____________, and ________, where $n$ is an odd integer.

  Examples: ___________________________________
- A geometric topic that often generates quadratic equations is __________________ ___________________.

- In a right triangle, one of the angles is a ________________ angle (measures 90°), and the side opposite this angle (the longest side) is called the _________________. The other two sides are called __________________.

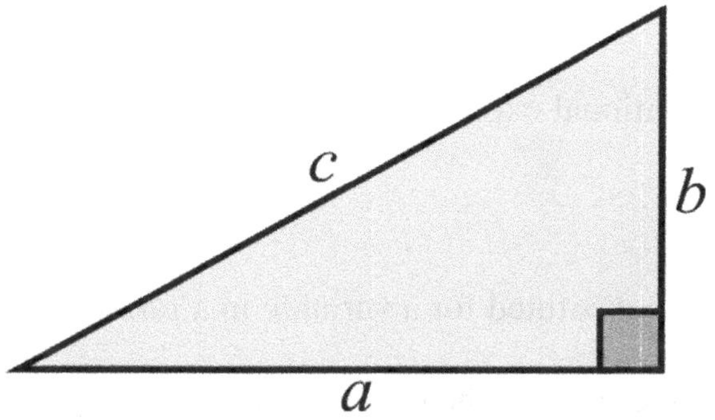

## Introduction to Rational Expressions

- A ________________ number is the technical name for a fraction in which both the numerator and denominator are integers.
- A ________________ expression is the technical name for a fraction in which both the numerator and denominator are polynomials.
- The denominator of a rational expression can ________________ be 0. Division by 0 is undefined.
- If a numerical value is substituted for a variable in a rational expression and the denominator assumes a value of 0, we say that the expression is ______________ for that value of the variable. (These values are called ______________ on the variable.)
- You can find restrictions on the variable in a rational expression by setting the denominator to zero.
    - Example:

$$\frac{7}{2x-5}$$

$$2x-5=0$$

$$2x=5$$

$$x=\frac{2}{5}$$

This means that any other number other than $\frac{2}{5}$ can be substituted for x. The restriction would be written as: $x \neq \frac{2}{5}$

- If the numerator of a rational expression has a value of 0 and the denominator is not 0 for that value of the variable, then the expression is defined and has a value of 0.If both numerator and denominator are 0, then the expression is undefined just as in the case where only the denominator is 0.

- A fraction (or rational number) is a number that can be written in the form $\frac{a}{b}$ where $a$ and $b$ are integers and $b \neq 0$. (Remember, no denominator can be 0.)

- Review the list of arithmetic rules for rational numbers below:

  - The ______________ Principle: $\frac{a}{b} = \frac{a \cdot k}{b \cdot k}$ $where\, b, k \neq 0$

  - The ______________ of $\frac{a}{b}$ $is$ $\frac{b}{a}$ and $\frac{a}{b} \cdot \frac{b}{a} = 1$ $where\, a, b \neq 0$.

  - ________________: $\frac{a}{b} \cdot \frac{c}{d}$ $where\, b, c, d \neq 0$

  - ________________: $\frac{a}{b} \div \frac{c}{d} = \frac{a}{b} \cdot \frac{d}{c}$ $where\, b, c, d \neq 0$

  - ________________: $\frac{a}{b} + \frac{c}{b} = \frac{a+c}{b}$ $where\, b \neq 0$

  - ________________: $\frac{a}{b} - \frac{c}{b} = \frac{a-c}{b}$ $where\, b \neq 0$

- The fundamental principle can be used to ______________ (or simplify) a rational expression to lower terms, for multiplication or division and to build a rational expression to higher terms, for addition or subtraction.

- A ________________ expression is said to be reduced to lowest terms if the numerator and denominator have no ____________ factors other than 1 and −1.

## Multiplication and Division with Rational Expressions

- To multiply any two (or more) rational expressions, you will need to:

  1. Completely ______________ each numerator and denominator.

  2. __________________ the numerators and multiply the denominators, keeping the expressions in factored form.

  3. "__________________ __________________" any common factors from the numerators and denominators. Remember that no denominator can have a value of __________________.

- If $P$, $Q$, $R$, and $S$ are ______________ and $Q, S \neq 0$, then $\frac{P}{Q} \cdot \frac{R}{S} = \frac{P \cdot R}{Q \cdot S}$.

- Generally, the ____________ will be left in factored form and the _____________ multiplied out.

- To divide any two rational expressions, ________________ the first fraction by the ____________ of the second fraction.

- If $P, Q$, $R$, and $S$ are polynomials with $Q, R, S \neq 0$, then $\frac{P}{Q} \div \frac{R}{S} = \frac{P}{Q} \cdot \frac{S}{R}$. Notice that $\frac{S}{R}$ is the __________________ of $\frac{R}{S}$.

## Addition and Subtraction with Rational Expressions

- To add rational expressions with a common denominator, proceed just as with fractions: add the _______________ and keep the common _________________. Reduce if possible.

- If you use the entire expression in the numerator (or denominator) to reduce, we are left with a factor of _______________.

- An example of adding polynomials is below. Fill in the missing parts:

$$\frac{1}{x^2+7x+10}+\frac{2x+3}{x^2+7x+10}$$

$$\frac{1+2x+3}{x^2+7x+10}$$

$$\frac{\phantom{xxxx}}{x^2+7x+10}$$

$$\frac{2(x+2)}{(\quad)(\quad)}$$

$$\frac{2}{\phantom{xx}}$$

- When the denominators are not the same for each polynomial you are adding or subtracting, you will need to find the _______________ for the polynomials by:

  1. Completely ______________ each polynomial (including prime factors for numerical factors).

  2. Form the ________________ of all factors that appear, using each factor the most number of times it appears in any one polynomial.

- The LCM of a set of denominators is called the least common _______________ (LCD).

- The steps for adding rational expressions with different denominators are:

1. Find the LCD (the LCM of the ________________).

2. Rewrite each fraction in an ______________ form with the LCD as the denominator.

3. Add the numerators and keep the common _____________.

4. _________________ if possible.

- When working with a ____________ sign and fractions you can assign the negative sign in any of the following ways: $-\frac{P}{Q} = \frac{P}{-Q} = \frac{-P}{Q}$

- A common mistake when subtracting rational expressions by not subtracting the _________________ numerator. Make sure you subtract the entire numerator in the second polynomial from the first. Group the numerator in the second polynomial with parenthesis.

# Complex Fractions

- A ________________ ________________ is a fraction in which the numerator and/or denominator are themselves fractions or the sum or difference of fractions.

- Below is an example of a complex fraction. Rewrite this fraction using grouping symbols and fractions:

$$\frac{\frac{1}{x+3}-\frac{1}{x}}{1+\frac{3}{z}}=\left(\frac{\quad}{\quad}-\frac{\quad}{\quad}\right)\div\left(\quad+\frac{\quad}{\quad}\right)$$

- A complex fraction indicates that the numerator is to be ______________ by the denominator.

- There are two ways to simplify complex fractions. One method is:

  1. Simplify the numerator so that it is a ____________ rational expression.

  2. Simplify the denominator so that it is a ________________ rational expression.

  3. __________________ the numerator by the denominator and reduce to lowest terms.

- Another way to simplify complex fractions is to:

  1. Find the ________________ of all the denominators in the numerator and denominator of the complex fraction.

  2. Multiply both the numerator and denominator of the complex fraction by this ____________.

3. ________________ both the numerator and denominator and reduce to lowest terms.

- A complex algebraic ____________ is an expression that involves rational expressions and more than one operation. It is simplified using the rules for _________________________.

# Division by a Monomial

- Fractions in which the numerator and denominator are polynomials are called

  ________________ ________________. (No denominator can be ______________.)

- The ______________ of fractions with the same denominator can be written as a

  ___________ fraction by adding the numerators and using the common denominator.

- If instead of adding the fractions, you should start with the sum and want to

  ________________ the numerator by the denominator (with a monomial in the

  denominator), you will divide each term in the numerator by the monomial denominator and

  ________________ each fraction.

## The Division Algorithm

- The ______________ ________________ (called long division) is the process (or series of

  steps) that we follow when dividing two numbers.

- Write both polynomials in order of _____________ powers. If any powers are missing, fill

  in with ___________.

- Simplify the problem below by using long division and write down an explanation for

  each step.

| Problem to evaluate. | Explanation of each step. |
|---|---|
| $\frac{6x^2-7x-2}{2x-1}$ | |
| $2x-1\overline{)6x^2-7x-2}$ | |
| $2x-1\overline{)6x^2-7x-2}$ | |
| $2x-1\overline{)6x^2-7x-2}$ | |
| $2x-1\overline{)6x^2-7x-2}$ | |

| | |
|---|---|
| $2x-1\overline{)6x^2-7x-2}$ | |
| $2x-1\overline{)6x^2-7x-2}$ | |
| $2x-1\overline{)6x^2-7x-2}$ | |
| $2x-1\overline{)6x^2-7x-2}$ | |

## Solving Equations with Rational Expressions

- A ratio is a comparison of two numbers by ____________.
- Ratios can be written in three forms:

  1.

  2.

  3.

- A proportion is a special type of equation stating that two ____________ are equal.
- Proportions can be used to solve many everyday types of ____________ ________________. Using the correct ____________ of units on both sides of the proportion is critical. One of the following conditions must be true:

  1. The ____________ agree in type and the denominators agree in type.
  2. The numerators correspond and the ____________ correspond.

- Steps to solving equations containing rational expressions are:

  1. Find the ________________ of the fractions.
  2. Multiply both sides of the equation by this ____________ and simplify.
  3. Solve the resulting equation. (This equation will have only ___________ on both sides.)
  4. Check each ____________ in the original equation. (Remember that no denominator can be 0 and any solution that gives a 0 denominator is to be discarded.)

- ________________solutions or extraneous roots are solutions given when solving the equation but are not true solutions to the particular problem.

## Applications: Rational Expressions

- Steps to solving a word problem with rational expressions are:

  1. Read the problem ________________. Read it several times if necessary.
  2. Decide what is asked for and assign a ___________________ to the unknown quantity.
  3. Draw a ___________________ or set up a ________________ whenever possible as a visual aid.
  4. Form an __________________ that relates the information provided.
  5. ___________________ the equation.
  6. Check your solution with the wording of the problem to be sure it makes ___________________.

- Problems involving _______________ usually translate into equations involving rational expressions.
- The basic idea is to represent what part of the work is done in _________________ unit of time.
- As in other lessons with application problems, use _______________to help you organize the information from the problem and your work.

# Evaluating Radicals

- ________________ ________________ is an expansion of the real number system and includes a group of numbers called imaginary numbers.
- Imaginary numbers are particularly useful in electrical ______________ and hydrodynamics.
- The symbol $\sqrt{\ }$ is called a ____________ sign.
- The number under the radical sign is called the ______________.
- The complete expression, such as $\sqrt{81}$, is called a radical or radical ________________.
- Square roots of ______________ numbers are not real numbers. For example, −4−−−√ is not a real number. There is no real number whose square is −4.
- A ______________ square root (or the principal square root) is represented with the $\sqrt{\ }$ symbol and $-\sqrt{\ }$ represents the ___________ square root.
- An ______________ number and a decimal representation is an infinite nonrepeating decimal.
- In the cube root expression $\sqrt[3]{a}$, the number 3 is called the index.
- In a square root expression such as $\sqrt{a}$, the index is understood to be ________________ and is not ______________.
- Expressions with square roots and cube roots (as well as other roots) are called ______________________ ___________.

## Simplifying Radicals

- Assuming $a$ and $b$ are positive real numbers, the properties of square roots are:

  1. $\sqrt{ab} =$ __________

  2. $\sqrt{\frac{a}{b}} =$ __________

- A square root is considered to be in simplest form when the radicand has no ____________ ________________ as a factor.

- If $x$ is a real number, then $\sqrt{x^2} =$ _______________. However, if we assume that the variable represents only __________________, __________________ numbers or 0 the need for absolute value signs is eliminated.

- A cube root is considered to be in simplest form when the radicand has no _____________ ________________ as a factor.

## Rational Exponents

- Complete the table describing exponents:

| Type of Root | Radical Notation and Exponential Notation |
|---|---|
| For square roots | |
| For cubed roots | |
| For fourth roots | |
| For $n^{th}$ roots | |

- If $n$ is a positive integer, then $\sqrt[n]{a} = a^{\frac{1}{n}}$ (assuming $\sqrt[n]{a}$ is a ____________number).

- The expression $\sqrt[n]{a}$ is called a _______________.

- The symbol $\sqrt{\phantom{a}}$ is called a _____________ _________________.

- *In* $\sqrt[n]{a}$, $n$ is called the ___________.

- *In* $\sqrt[n]{a}$, $a$ is called the ___________.

- If no index is given, it is understood to be ___________________. For example, $\sqrt{3} = \sqrt[2]{3} = 3^{\frac{1}{2}}$

- For the expression $\sqrt[n]{a}$ *or* $a^{\frac{1}{n}}$ to be a real number:

  1. when $a$ is nonnegative, $n$ can be any ___________, and

  2. when $a$ is negative, $n$ must be odd. (If $a$ is negative and $n$ is even, then $\sqrt[n]{a}$ is ___________.)

| Summary of the Rules for Exponents | |
|---|---|
| For nonzero real numbers a and b and rational numbers m and n, | Write a description or rule for each item in the left column. |
| The exponent 1 | |
| The exponent 0 | |
| The product rule | |
| The quotient rule | |
| Negative exponents | |
| Power rule | |
| Power of a product | |
| Power of a quotient | |

- If n is a positive integer, m is any ______________, and $a^{\frac{1}{n}}$ is a real number, then $a^{\frac{m}{n}} = (a^{\frac{1}{n}})^m = (a^m)^{\frac{1}{n}}$

- In ______________ notation: $a^{\frac{m}{n}} = (\sqrt[n]{a})^m = \sqrt[n]{a^m}$

- When converting from radical notation to exponential notation, you will simplify each expression by first changing it into an equivalent ______________ with ______________ ______________. Then rewrite the answer in simplified ______________ form.

## Addition and Subtraction with Radicals

- Like terms occur when each term contains the same ___________ ________________
- Write three expressions with like radicals, meaning each term contains the same

  ___________ expression:

  1.

  2.

  3.

- Simplify radicals before trying to ______________ or ______________the terms
- According to Learn, simplify the example: $\sqrt{32x} + \sqrt{18x}$

  1. Step One:

  2. Step Two:

  3. Step Three:

  4. Step Four:

## Multiplication with Radicals

- To find a product treat the two expressions as two ______________ and ______________ just as with polynomials.
- Using the ________________method, write down the steps for the example:

  $(\sqrt{3}+5)(\sqrt{3}-7)$ :

  1. Step One

  2. Step Two

  3. Step Three

  4. Step Four

## Rationalizing Denominators

- Calculations of sums and differences are much easier if the denominators are ________________ expressions.
- The objective is to find an equivalent fraction that has a rational number or an expression with no ________________ for a ____________.
- How to rationalize a denominator containing a square root or cube root:

   1. If the denominator contains a square root, multiply both the numerator and denominator by an expression that will give a ________________ with no ___________ ________________.

   2. If the denominator contains a cube root, multiply both the numerator and denominator by an expression that will give a denominator with no ________________ _____ ___________.

- What to if you need to rationalize a denominator containing a sum or difference involving square roots:

   - If the denominator of a fraction contains a sum or difference involving a square root, rationalize the denominator by multiplying both the numerator and denominator by the ______________ of the denominator. This is discussed below:

   1. If the denominator is of the form $a-b$ , multiply both the numerator and denominator by ___________ .

   2. If the denominator is of the form $a+b$ , multiply both the numerator and denominator by ______________ .

- The new denominator will be the ________________ of two squares and therefore will not contain a radical term.

# Equations with Radicals

- Write three examples of equations with radicals (involves at least one radical expression):

  1.

  2.

  3.

- If the radicals are square roots, we solve by ____________ both sides of the equations.
- When both sides of an equation are raised to a power, an _____________ solution may be introduced. Be sure to check all solutions in the original equation.
- Steps on solving radical equations are:

  1. I_________________ one of the radicals on one side of the equation. (An equation may have more than _____________ radical.)

  2. Raise both sides of the equation to the power corresponding to the ____________ of the radical.

  3. If the equation still contains a radical, repeat steps 1 and 2.

  4. Solve the equation after all the radicals have been _____________.

  5. Be sure to check all possible solutions in the original equation and ____________ any extraneous solutions.

## Complex Numbers

- Complex numbers include all the ______________ numbers and the even roots of ________________ numbers.

- The first step in the development of complex numbers is to define ____________.

- If $a$ is a positive real number, then $\sqrt{-a} = \sqrt{a} \cdot \sqrt{-1} = \sqrt{a} \cdot i$ . Please notice that the number $i$ is _______________ under the radical sign. To avoid confusion, we sometimes write $i\sqrt{a}$ .

  Example: $\sqrt{-144} = \sqrt{-1} \cdot \sqrt{144} = i \cdot 12 = 12i$

- The standard form of a complex number is $a+bi$, where $a$ and $b$ are real numbers. $a$ is called the real part and $b$ is called the imaginary part.

  - If $b=0$, then $a+bi=a+0i=a$ is a ____________ number.
  - If $a=0$, then $a+bi=0+bi=bi$ is called a ____________ _________________ number (or an imaginary number).

- Circle the real part and put a square around the imaginary part of the complex number below:

$$a+bi$$

- The following table illustrates the relationships among the various types of numbers you have studied. Please write examples of each type of number listed in the table.

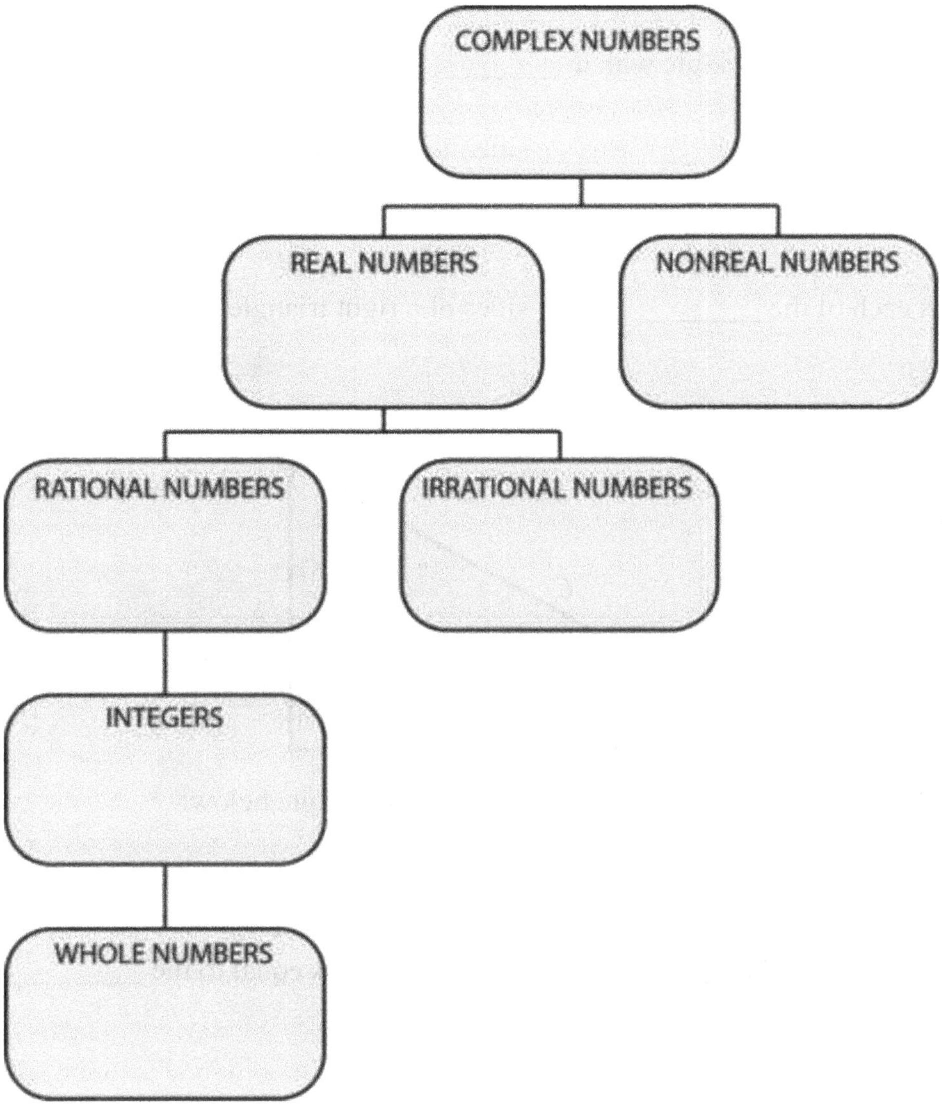


- When finding ____________ values, complex numbers $a+bi$ and $c+di$, if $a+bi=c+di$, then $a=c$ and $b=d$.
- When ____________ or ______________ complex numbers $a+bi$ and $c+di$, $(a+bi)+(c+di)=(a+c)+(b+d)i$ and $(a+bi)-(c+di)=(a-c)+(b-d)i$.

## Applications: The Pythagorean Theorem

- A right triangle is a triangle with a ______________ angle (____________ degrees).
- A hypotenuse is the ______________ side of a right triangle (the side ________________ the right angle).
- A leg is each of the ______________ sides of a right triangle.

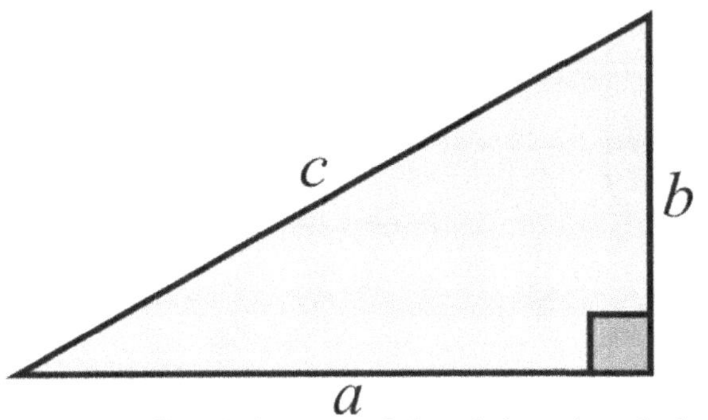


- Please label the parts (defined above) of the right triangle below:

- In a right triangle, the square of the _____________ is equal to the __________________ of the squares of the two legs:

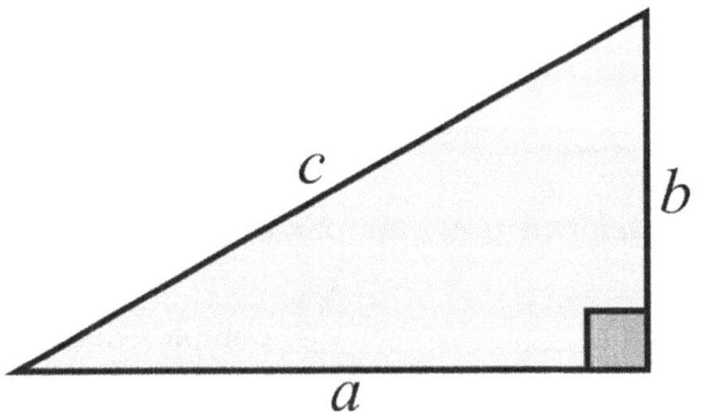


The Pythagorean Theorem is $a^2 + b^2 = c^2$, where $a$ and $b$ are ________________and $c$ is the __________________.

## Applications: Distance Formula and Circles

- In a right triangle, if $c$ is the length of the _______________ and $a$ and $b$ are the lengths of the _______________, then $a^2 + b^2 = c^2$.
- Remember, the hypotenuse is the side _________________ the right angle and is the longest side.
- You can calculate the distance between two points by creating a right triangle and solving for the _______________ of the triangle.
- For two points $P(x_1, y_1)$ and $Q(x_2, y_2)$ in a plane, the _________________ between the points is $d = \sqrt{(x_2 - x_1)^2 + (y_2 - y_1)^2}$.
- The formula for the ___________________ between the two points $P(x_1, y_1)$ and $Q(x_2, y_2)$ is $\left(\frac{x_1 + x_2}{2}, \frac{y_1 + y_2}{2}\right)$.
- A circle is the set of all points in a plane that are a fixed _______________ from a fixed point.
- The fixed point is called the _____________ of the circle.
- The distance from the center to any point on the circle is called the __________________ of the circle.
- The distance from one point on the circle to another point on the circle measured through the center is called the _______________ of the circle.
- The diameter is twice the length of the _________________.

- Please label the parts of the circle (defined above) in the image below:

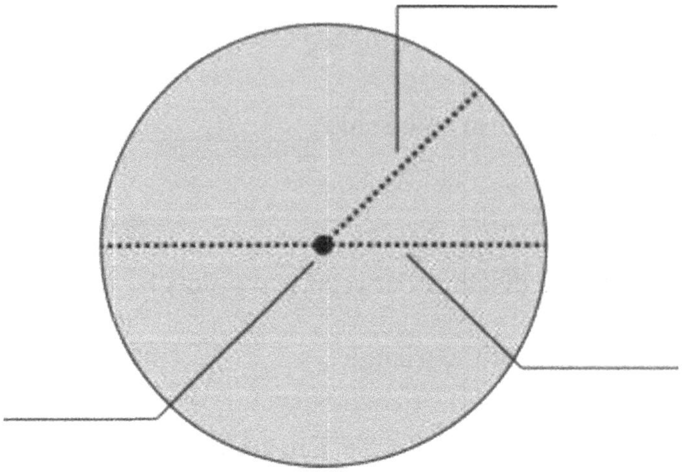

- The equation of a circle with ____________ $r$ and center at $(h, k)$ is $(x-h)^2+(y-k)^2=r^2$.

- Please label the parts of the graph below:

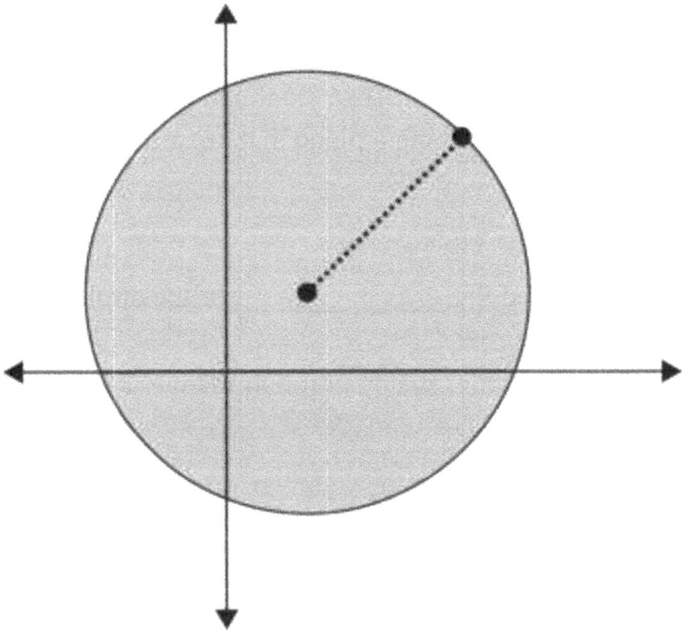

- If the center is at the origin, (_______, _______), the equation simplifies to $x^2+y^2=r^2$.

## Functions and Function Notation

- The ordered pairs discussed in this text are ordered pairs of real numbers. However, more generally, ordered pairs might be other types of pairs such as:

  1.

  2.

  3.

- A ________________ is a set of ordered pairs of real numbers.
- The ______________, $D$, of a relation is the set of all first coordinates in the relation.
- The ______________, $R$, of a relation is the set of all second coordinates in the relation.
- The domain consists of the set of ___-values for all points on the graph.
- The range consists of the set of _____-values for all points on the graph.
- A ________________ is a relation in which each domain element has exactly _________ corresponding range element.
- The definition can also be stated in the following ways:

  1. A function is a relation in which each first coordinate appears only ____________.
  2. A function is a relation in which no two ordered pairs have the same ___________ coordinate.

- When using the ______________ ______________ test, if any vertical line intersects the graph of a relation at more than one point, then the relation is not a ______________.

- A ________________ function is a function represented by an equation of the form

$$y = mx + b.$$

- The ______________ of a linear function is the set of all real numbers: $D = (-\infty, \infty)$.

- In determining the domain of a function, one fact to remember is that no denominator can equal ___________.

- In ________________ notation, instead of writing __________ we write $f(x)$, which is read "_____________."

## Functions with Radicals

- A _______________ is a set of ordered pairs of real numbers.
- The _______________, D, of a relation is the set of all first coordinates in the relation.
- The _______________, R, of a relation is the set of all second coordinates in the relation.
- A function is a _______________ in which each domain element has exactly one corresponding range element.
- If any vertical line intersects the graph of a relation at more than _______________ point, then the relation is not a function.
- A radical function is a function of the form $y=\sqrt[n]{g(x)}$ in which the radicand contains a _____________ expression.
- The domain of such a function depends on the _______________, *n*:
  - If *n* is an _______________ number, the domain is the set of all *x* such that $g(x)\geq 0$.
  - If *n* is an _______________ number, the domain is the set of all real numbers, $(-\infty,\infty)$.
- Find the domain of the radical function $f(x)=\sqrt{2x+3}$. Keep in mind that because the index is 2, the radicand must be nonnegative. (That is, the expression under the radical sign cannot be negative.)

  1.

  2.

  3.

- Find the domain of the radical function $y = \sqrt{x-7}$. Notice that the index is 3, an odd number. Because of this, the domain is ______________________________ or (_________, ____________)

- To solve a radical function, use a ________________ to help you organize your work.

## Quadratic Equations: The Square Root Method

- ________________ equations are equations that can be written in the form $ax^2+bx+c=0$ where $a$, $b$, and $c$ are real numbers and $a \neq 0$.

- Steps to solving a quadratic equation by factoring are:

  1. ________________ or ______________ terms as necessary so that 0 is on one side of the equation and the equation is in the standard form $ax^2+bx+c=0$ where $a$, $b$, and $c$ are real numbers and $a \neq 0$.

  2. ________________ completely. (If there are any fractional coefficients, multiply each term by the least common denominator first so that all coefficients will be integers.)

  3. Set each nonconstant factor equal to ___________ and solve each linear equation for the unknown.

  4. Check each ______________, one at a time, in the original equation.

- The ______________ ________________ Property states that if:
  - $x^2=c$, then $x=\pm\sqrt{c}$
  - $(x-a)^2=c$, then $x-a=\pm\sqrt{c}$, or in other words $x=a\pm\sqrt{c}$.

- Solve the following quadratic equation using the square root property:

$(y+4)^2=8$

1.

2.

3.

# Quadratic Equations: Completing the Square

- In order to solve a quadratic equation by completing the square, you will need to:

  1. If necessary, divide or multiply on both sides of the equation so that the leading coefficient (the coefficient of $x^2$) is _______.

  2. If necessary, _____________ the constant term on one side of the equation.

  3. Find the ________________ that completes the square of the polynomial and add this constant to both sides. Rewrite the polynomial as the square of a binomial.

  4. Use the square root property to find the _______________ of the equation.

## The Quadratic Formula

- The quadratic formula gives the ________________ of any quadratic equation in terms of the coefficients *a*, *b*, and *c* of the general quadratic equation.
- To develop the quadratic formula, we solve the general quadratic equation by ________________ the ______________.
- The ______________ ________________ is $x = \frac{-b \pm \sqrt{b^2 - 4ac}}{2a}$ when $a \neq 0$.
- The quadratic formula should be ________________.
- Solve the quadratic equation $7x^2 - 2x + 1 = 0$ by using the quadratic formula:

  Step 1:

  Step 2:

  Step 3:

  Step 4:

  Step 5:

  Step 6:
- Avoid making a mistake of simplifying fractions by dividing the denominator into only ________________ of the terms in the numerator.
- The expression $b^2 - 4ac$, the part of the quadratic formula that lies under the radical sign, is called the ________________.
- The discriminant identifies the ________________ and ______________ of solutions to a quadratic equation.
- There are three possibilities: the discriminant is either:

1.

2.

3.

- Complete the table below describing the different types of discriminants and the nature of solutions:

| Discriminant | Nature of Solutions |
|---|---|
| | Two real solutions |
| $b^2 - 4ac = 0$ | |
| | Two nonreal solutions |

## Applications: Quadratic Equations

- Strategies for Solving Word Problems:

1. Understand the ________________.

    - ________________ the problem carefully. (Read it several times if necessary.)
    - If it helps, ______________ the problem in your own words.

2. Devise a ________________.

    - Decide what is asked for; assign a variable to the unknown quantity. Label this ______________ so you know exactly what it represents.
    - Draw a ________________ or set up a ______________ whenever possible.
    - Write an ________________ that relates the information provided.

3. Carry out the ____________.

    - Study your picture or diagram for insight into the ______________.
    - Solve the ________________.

4. Look ____________ over the results.

    - Does your solution make ______________ in terms of the wording of the problem?
    - Check your solution in the equation.

- In a right triangle, if c is the length of the hypotenuse and a and b are the lengths of the legs, then $a^2 + b^2 = c^2$, which is known as the ______________ __________________ .

## Quadratic Functions

- The ______________ ____________________ ____________________ shows that if any vertical

  line intersects a graph in more than one point, then the relation is not a function.

- Any function that can be written in the form $y = ax^2 + bx + c$ where *a, b, and c* are real numbers and $a \neq 0$ is a __________________ ____________________.

- The ________________ is the set of all real numbers. That is, $x$ can be replaced by any real number and there will be one corresponding $y$-value.

- Characteristics of quadratic functions of the form $y = ax^2$:

  - The ________________ is at (0, 0).
  - If $a > 0$, the parabola "opens _______."
  - If $a < 0$, the parabola "opens _____________."
  - The bigger $|a|$ is, the _____________ the opening.
  - The smaller $|a|$ is, the ____________________ the opening.
  - The line $x = 0$ (the $y$-axis) is the ________________ of ________________.

- The ________________ of a vertical parabola is either the lowest point or the highest point on the parabola.

- For a parabola with its equation in the form $y = a(x-h)^2 + k$:

  - If $a > 0$, then the parabola opens upward and ($h$, $k$) is the lowest point and the $y$-value $k$ is called the _____________ value of the function.
  - If $a < 0$, then the parabola opens downward and ($h$, $k$) is the highest point and $y$-value $k$ is called the ____________________ value of the function.